KU-350-481

OCEANLIFE

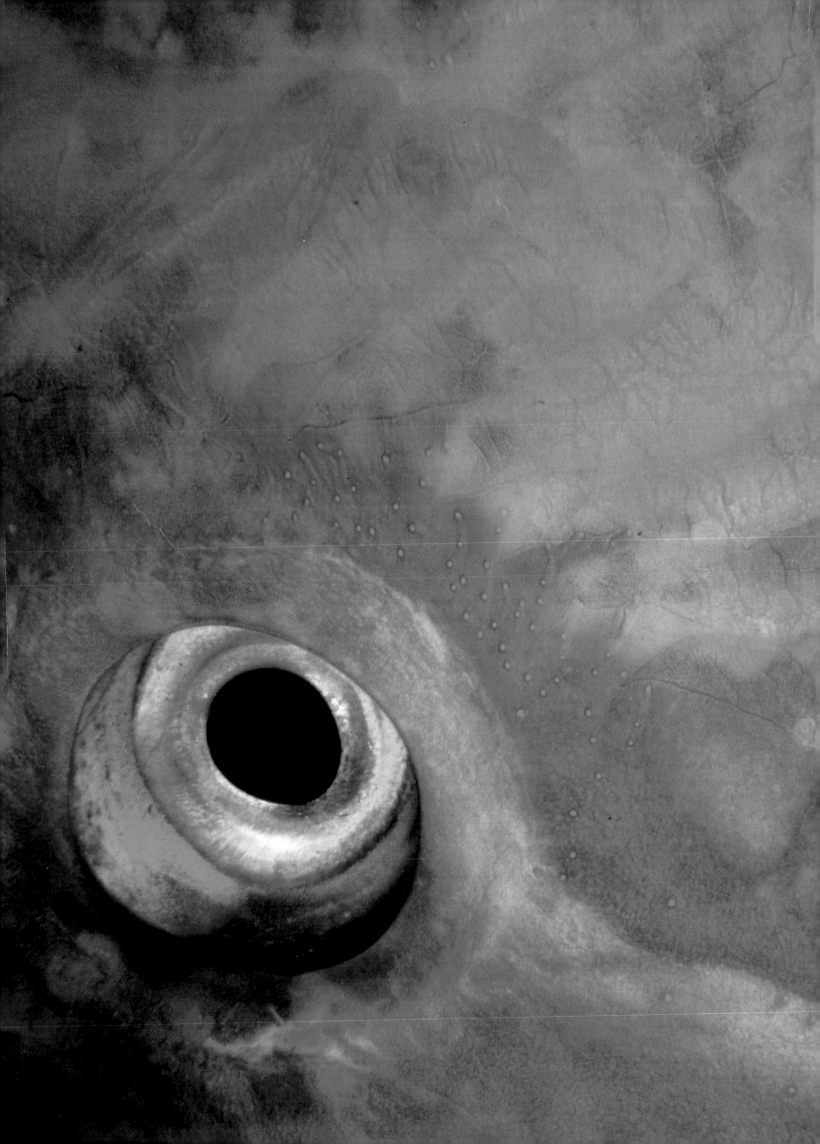

Metriaclima greshakei "Ice Blue Zebra"

 Male

 Female

- Things you need to know about these amazing fish:

- They are three generations removed from Lake Malawi, tank raised in Nairn water!

- In order to effectively keep these fish you will need a tank 36" or larger, with good filtration and filled with rocks. They aggressively eat most plants, but some success can be found with Java Ferns, Java Moss or Vallis.

- These fish require a PH of 7.5 or higher. Highland water is notoriously soft, coming out of the tap at around 6.8-7.3. You can raise the PH level of your water by using chemical buffers such as "PH up" and maintain a high PH with calcareous rocks like limestone or by using coral gravel on the tank floor.

- Greshakei (like all Zebras) are best kept with 1 male to 3 females, you can pretty much guarantee babies if you keep this ratio. It's also the best way to see the males brilliant colours.

- Fully grown males will reach 5 to 6 inches and females around 4-5" They need to be fed a mixed diet of quality cichlid flakes, algae pellets or spiriluna wafers, all of which are available in store.

- These fish are best kept with other Lake Malawi Mbuna as they can be very aggressive!

OCEANLIFE

Sally Morgan and Pauline Lalor

PARKGATE
BOOKS

First published in 2000 by
PRC Publishing Ltd,
Kiln House, 210 New Kings Road, London SW6 4NZ

This edition published in 2000 by
Parkgate Books
London House
Great Eastern Wharf
Parkgate Road
London
SW11 4NQ

© 2000PRC Publishing Ltd

All rights reserved. No part of this publication may
be reproduced, stored in a retrieval system, or transmitted
in any form or by any means, electronic, mechanical,
photocopying, recording, or otherwise, without the prior
written permission of the Publisher and copyright holders.

British Library Cataloguing in Publication Data:
A catalogue record for this book is available from the British Library.

ISBN 1 90261 666 6

Printed and bound in China

Cover: Gray snappers with bluestriped, ceasar and French grunts.

Page 2: The eye of a parrot fish. See page 50.

Above: A white, soft coral. See page 271.

CONTENTS

INTRODUCTION

Seen from space, Earth is a blue planet. Oceans cover nearly three-quarters of its surface and plunge to depths of six kilometres (3.5 miles) or more. Compared with land, this watery world is hardly explored, yet it is more complex than any land-based ecosystem—the oceans of the world are teeming with life. The sea's intricate food webs support more life by weight and a greater diversity of animals than anywhere else on the planet, from sulphur-eating bacteria clustered around deep-sea vents to fish that light up like a neon sign to lure their prey. From the shallows of the shoreline, where the world's vast seas meet the land, to the deepest depths of mighty canyons on the sea floors, nature has managed to fill every possible niche with a creature admirably suited to their watery home. From enormous and graceful whales to the tiniest, passively floating micro-organisms, the rich variety of life beneath the seas is almost unimaginable; yet to many of us land-dwelling humans our planet is Earth, which could be seen to imply that it is exclusively composed of soil, in which plants grow and on which animals live. Dry land seems to many to be the norm and indeed, strange as it may seem in this day and age, when travel has never been easier, there are still people living in parts of the world who have never seen the ocean.

Left: Earth is the only planet in the solar system that has liquid water. Conditions are too hot on Venus and Mercury and too cold on the outer planets.

Right: Bubbles rise to the surface from an area of volcanic activity on the sea floor.

However, for thousands of years many other people have lived their lives entwined with the sea, harvesting its bounties and struggling to survive the cruelty that the ocean can sometimes mete out, always trying to second-guess the sea to safeguard their own lives and capitalise on the riches below their fragile boats. More recently, scientists have attempted to complete the human understanding of the mysterious deeps, which continue to hold secrets that are difficult to unravel. But even with all of our modern technology, there is still much that we do not know of the oceans and, unlike the land, many places below the waves that humans have never seen. Indeed, we know so little of the deepest parts of our oceans that they may very well be home to huge sea creatures such as the giant squid. It is a strange fact that while humans send probes to the outer reaches of the galaxy there is still conjecture on the possibility of the existence of sea-monsters.

It is humbling to realise that life itself originally began in the water and slowly evolved into land forms. The highly evolved whales and dolphins that we all love so much are mammals that have returned to the sea and it seems, in our quest to explore the oceans, that we are following them.

World of Water

The sea floor is not a flat, featureless plain but is criss-crossed by folds and troughs that compete for ruggedness with the highest mountains. Not only do the oceans have canyons deep enough to hide the Himalayas, but they are also the setting for what is by far the largest geological feature on the planet—a single, 60,000 kilometre-long (37,000 mile-long) mountain range that snakes its way continuously through the Atlantic, Pacific, Indian and Arctic oceans, completely encircling the planet. The oceans are immense. They are important because they are the places in which life first arose and in which the greatest range of animal types still live and, as such, hold many of the answers to our own existence on the planet. The oceans also influence the Earth's climate and are a huge source of food. Millions of tonnes of marine life are harvested each year and many people continue to rely on the sea for their livelihoods.

It is the unique properties of water that allow it to support life. Water is a solvent, so sea water contains almost every chemical element in some form and quantity. There are 34.7 grammes (1.22 ounces) of mineral salts in every kilogram (2.2 pounds) of salt water but just six elements make up 90 per cent of all the minerals. These are chlorine, sodium, sulphur (as sulphate), magnesium, calcium and potassium. Other minerals are only present in tiny quantities but they are essential to marine life, especially the phytoplankton—the microscopic plants that live in the surface layer of the water. Oxygen dissolves in water and is available at all depths, so life can exist even in the deepest parts of the ocean. Water is also transparent, so light can penetrate the upper layers of the oceans, allowing plants near the surface to photosynthesise and make food as they do on land. Water has a high density, which allows organisms to float and a high thermal capacity, which means that it takes a lot of energy to heat up a large body of water, so the temperature remains constant for much of the year, with only the surface layer showing any significant variations.

In addition, water has a low viscosity, which means that wind can get a frictional grip on the surface and create waves and currents. The patterns of flow result from the interaction of winds and the rotation of the earth as it spins on its axis. In the Northern Hemisphere, there is a clockwise set of surface currents driven by the trade winds and prevailing westerlies. Circulating water has to be replaced, so counter-currents also form. At certain points on the Earth's surface, conditions such as high salinity and low temperatures combine to make the water dense. It sinks to the bottom and becomes part of the deep water circulation where cold water flows along the sea floor and may eventually mingle with the overlying mass of warm water. Some currents flow quite rapidly. The Gulf Stream, for example, moves at two metres per second (6.5 feet per second), but the deeper ones are often much slower, frequently less than 0.2 metres per second (six inches per second). These ocean currents affect the distribution of heat, oxygen, nutrients and even the dispersal of floating organisms such as plankton, spores and the seeds of plants and it is the mass movement of warm and cold water around the oceans that is largely responsible for the global climate. The Gulf Stream, for example, brings warm water from the tropics to Northern Europe, giving the western coast of Ireland and

Above: Dolphins playing in the sea.

Above Left: Islands are isolated by the open oceans.

Scotland a mild winter climate. A cold current flows up the west coast of South America and although this means the winters can be harsh, the cold waters of this area are highly productive and support rich fisheries.

The oceans also have vast reserves of commercially valuable minerals such as nickel, iron, manganese, copper and cobalt, while pharmaceutical and biotechnology companies are already financing research that is analysing deep-sea bacteria, fish and marine plants, in search of substances that they might someday turn into disease fighting drugs.

More than 97 per cent of the water on Earth is salt sea-water. Of the remaining three per cent that is fresh water, two thirds is locked up in glaciers and in the ice caps of the Arctic and Antarctic. While land-living plants and animals need fresh water in order to survive, marine organisms have adapted to living in salt water, but even so they face a number of problems. The water surrounding their bodies contains more concentrated salt than their tissues, and this means that water is continually leaving their bodies and they have to find ways to reduce the amount of water that is lost. Some achieve this by drinking salt water and excreting the salts back into

the ocean, while others have kidneys that work very slowly and do not produce much in the way of urine.

The oceans are a huge reservoir of water, which evaporates from the surface and is then transported through the air as vapour. When it reaches land it falls back to the ground as rain, snow or ice and this water drains into streams and rivers from where it is discharged back into the oceans to complete the cycle. The water cycle takes the water through the rocks and soils of the land and slowly, over millions of years, salt from the rocks is dissolved in the water and washed into the sea, creating the saltiness of the water.

Above: When water turns to ice, it becomes less dense. This is why ice floats on water.

Above Left: Light can penetrate water to depths of about 200 metres (650 feet).

Overleaf: This satellite image over North America shows swirling weather systems of cloud produced by the sun's rays drawing vapour from the oceans.

Above Left: Waves are created by wind moving over the surface of the ocean.

Top: The Arctic Ocean is the world's smallest ocean and it is surrounded by land. It is covered by ice for much of the year.

Left: Two-thirds of the fresh water on the planet is locked up in ice, and nine-tenths of this is found in the ice shelves of Antarctica, where the average depth of the ice is two kilometres (1.25 miles).

Above: The remaining ice is found in the world's glaciers such as the Harvard Glacier in Alaska.

Above: The water vapour falls back to earth as rain, snow or ice. It runs off the land into streams and rivers that carry the water back to the oceans.

Left: Water that evaporates from the surface of the ocean condenses to form cloud.

Right: The heavy rainfall on the plateau lands of Venezuela creates a torrent of water that tumbles over Angel Falls, the tallest waterfall in the world, on its way to the Atlantic Ocean.

Early Ocean Life

But what of life? Where did it begin? Clues are to be found on a beach in Australia. At Shark Bay, a carpet of strange stony cushions, called stromalites, is revealed at low tide. Though it is hard to believe, these cushions are actually living. They are found in Shark Bay because of the unique conditions there. The movement of water in and out of the bay is slowed by a sand bar and during the day the hot sun evaporates some of the water, creating pools of concentrated salty water. Few animals can survive in these pools, so mats of blue-green algae, known as cyanobacteria, are built up. Cyanobacteria are simple, singled-celled organisms that use sunlight to make their own food, just like plants. As the cyanobacteria grow, they secrete lime to form the stony cushion. These primitive organisms are very similar to those that lived 3,000 million years ago. Standing beside these cushions you can feel a little of how the world

Above: Volcanic eruptions were common 3,000 million years ago.

Above Left: Low tide in Shark's Bay reveals the strange-looking living boulders called stromalites.

Far Left: Cyanobacteria were once called blue-green algae because of their colour.

Left: Diatoms make up much of the plankton in the oceans.

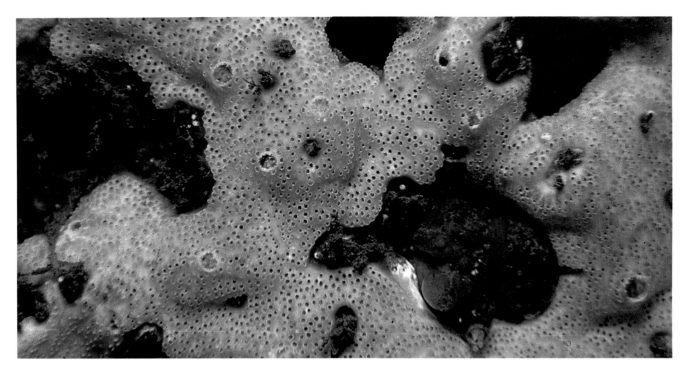

Top: Sponges are filter feeders, drawing in water through their pores.

Above: This vase sponge secretes tiny needles of silica or lime, which form an elaborate and intricate scaffold around its cells.

Right: Hundreds of tentacles—each covered in sting cells—hang beneath this jellyfish.

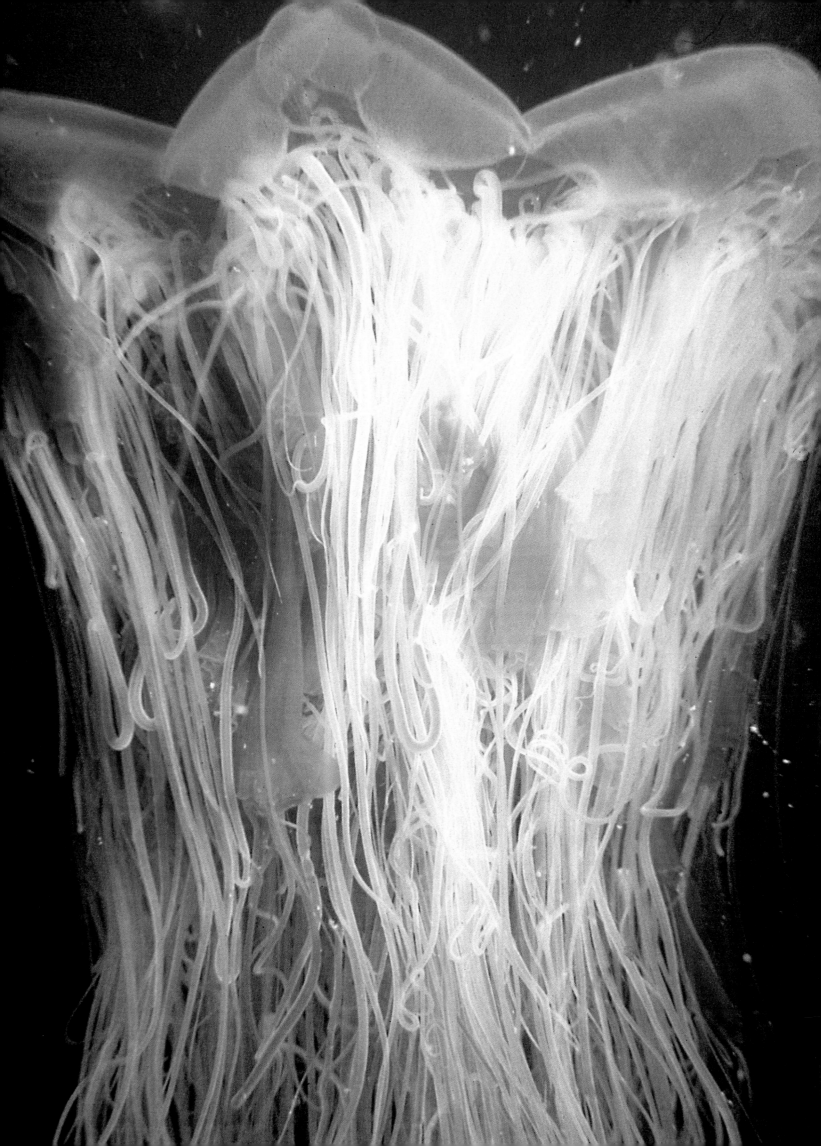

must have been at the time when life itself was being created and the planet looked very different. At this time of frequent volcanic eruptions and violent electrical storms, the seas were newly-formed from massive clouds of water vapour that had cooled down and condensed while the atmosphere was very thin and made up of hydrogen, carbon monoxide and methane. There was no oxygen.

The very first life forms were bacteria; some of which fed on organic carbon while others evolved the ability to use light as a source of energy. It was these forerunners of plants that were responsible for changing the planet's atmosphere, for they were producers of oxygen. Over millions of years these tiny organisms slowly increased the level of oxygen in the atmosphere, and with this fundamental change came new animals that depended on oxygen. One important form of oxygen, ozone, formed a protective layer high in the atmosphere, screening the planet from harmful ultraviolet light.

The first animals appeared more than 1,000 million years ago. They were simple, single-celled creatures called protozoans that lived in water—even today, there are many thousands of different types of protozoans living in both fresh and salt water—and the

next step in the evolutionary process was when groups of cells started to combine to form a colony. Approximately 800 million years ago, the first sponges appeared. These were (and still are) soft shapeless organisms found on the seabed. Covered with tiny pores they behave like miniature vacuum cleaners, drawing in water through the small pores and expelling it through larger ones. As the water passes through, tiny particles are filtered from the water. Sponges have to process vast quantities of water in order to get enough food—a sponge the size of a teacup can filter 5,000 litres (1,300 gallons) of water a day. Some sponges surround their cells with a flexible substance that acts as a skeleton. When the sponge is boiled, the skeleton is left behind and this is what people used to use in their baths.

The first truly multi-cellular organisms were the cnidarians—the jellyfish, anemones and coral. These animals are made from two layers of cells and they all have sting cells consisting of a poisoned or barbed thread, coiled up ready for release. When food or a predator comes close, the cells discharge the threads which wrap around or stick to the animal. The same thing happens if you are unfortunate enough to be stung by a jellyfish—tiny barbs embed in your skin causing pain and a rash. One important member of this group is the coral, the tiny organisms that are responsible for building coral reefs. The coral can secrete a skeleton of limestone around its soft body and while each coral animal is small, they live together in large colonies all linked together as if they were one. These animals, working together, are master builders, producing giant structures such as the Great Barrier Reef.

Sea pens are related to the soft corals and are so named because they look just like a quill pen. These are weird organisms, shaped liked feathers that rise vertically from the seabed. At night they produce a eerie purple light.

Eight hundred million years ago the muddy seabeds around the continents fostered the evolution of animals that could survive in the mud by burrowing through it and feeding on dead and decaying matter washed from the land. The result was a group of animals known as worms. Some secreted a protective tube around their soft bodies. Their heads, bearing many tentacles, would emerge from the tube to filter the water for food particles. Today, there are many marine worms including the lugworm, which lives in muddy estuarine waters, and the ragworm, an active hunter of the sea floor.

Above: The sea anemone has a ring of tentacles around its mouth to catch food.

Above Left: A sea pen rises from the seabed.

Left: An orange gorgonian coral growing on a coral reef in the Solomon Islands.

Above Left: There are many remarkable forms of worms in our oceans. The fire worm is an active predator on the sea floor.

The feather duster worm (**Above**) and the peacock worm (**Left**) are both sessile, meaning that they are permanently attatched to the seabed. Both build a tube around their bodies and their tentacles filter out particles in the water.

There are many strange-looking fossils that let us glimpse the life that lived hundreds of millions of years ago. A few creatures have survived virtually unchanged until today but most have disappeared, living on only as fossils. About 500 million years ago, the seabed was alive with three groups of animals—the trilobites, with their segmented bodies that looked just like giant woodlice; the brachipods, tiny shelled animals; and the crinoids, stalked flower-like animals. These three were the evolutionary ancestors of some of the major groups of animals found in the oceans today.

The trilobites were a highly successful group that dominated the seas for almost 200 million years. Then, about 250 million years ago, they started to disappear, but one survived—the horseshoe crab. Today, this animal, with its domed shell and spiked tail, lives in deep Atlantic waters. Each spring, horseshoe crabs migrate to the shores of Maryland where, at full moon, they emerge from the sea in their thousands. The females lay their eggs and the males quickly fertilise them with their sperm. The eggs settle in the sand and, at the next high tide four weeks later, the eggs hatch and the larvae swim off into the ocean.

Brachipods evolved into modern day snails. These creatures secreted a shell of calcium around their body for protection and moved along the sea floor using a flat muscular structure called the foot. Today, there is a wide diversity of such molluscs including snails with single pyramidal-shaped shells, such as the limpet, coiled shells and paired shells. Snails have a tongue covered in tiny teeth, called a radula, which they use to rasp algae off rocks and, in some cases, to bore holes through other snail shells or even into rock.

A few snails have abandoned their shells entirely. These animals are the sea slugs or nudibranchs. They are a diverse and brightly coloured group of creatures, the top of their soft bodies being covered in waving tentacles, which are armed with sting cells taken from the bodies of the anemones and jellyfish they feed on.

Snails with two hinged shells are called bivalves—the most familiar of these being mussels and oysters. They remain sedentary on the seabed and filter the water to obtain food. Some, like the giant clam, can grow to great sizes.

The third group of molluscs is related to the huge coiled fossils called ammonites, which were a highly successful group. Giant, wheel-like fossils have been found embedded in limestone rock, giving us an idea of what they looked like. Each had a coiled shell, inside which was a series of air-filled spaces that gave the creature buoyancy in the water. As it grew larger, it added new chambers to give a little more buoyancy. One living relative of these animals still survives: the nautilus, which is related to the fast-moving members

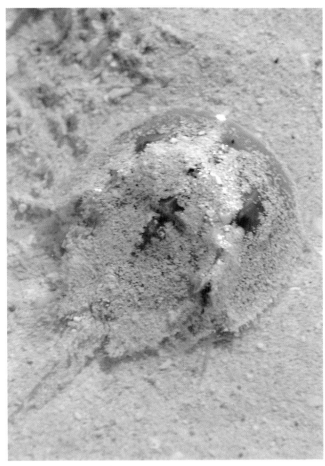

During the Cambrian period, 500 million years ago, both trilobites (**Above Left**) and brachiopods (**Above**) were abundant. The name "trilobite" refers to the furrows that divide the dorsal surface into three. Most were just a few centimetres long, but a few were more than 50 centimetres (20 inches) in length. Brachiopods were animals with two hinged shells, a bit like the mussels we know today.

Top Right: A horseshoe crab emerges from the sea.

Both topshells (**Top**) and conch (**Right**) are gastropods with a simple spiral shell.

of the group—the octopus, squid and cuttlefish. Squid and cuttlefish still have the remnants of a shell deep inside their body—people who own birds will be familiar with the cuttle bone which is a great source of calcium—but the octopus has no shell at all. The animals of this group are known as cephalopods, meaning "head-feet" animals. The squid can grow to lengths in excess of ten metres (32 feet) and there are many ancient stories and myths about giant squids that rise from the depths to pull ships under. Scars on some sperm whales, caused by the suckers of squid, suggest that animals of at least 13 metres (42 feet) exist and there may be still larger examples in the depths of the oceans.

The crinoids too, have survived through to this day and are yet another example of a living fossil. A crinoid has a long stalk-like structure that bears five arms at the top and are protected by plates of limestone embedded in their skin, which gives them a hard, spiny feel. These animals have given rise to the echinoderms of today—the spiny animals such as starfish, sea urchins and sea cucumbers—which have a fluid-filled skeleton known as a hydroskeleton. Water enters the animal through tiny openings in the sieve plate, or madreporite, on the upper surface and is drawn down into a ring canal from which radial canals fan out. In the case of a starfish, these service each arm. Each radial canal connects to short side branches with pairs of tube feet, each of which ends in a little sucker. By squeezing on the water within the system, the tube feet can be extended and contracted. Each tube foot is weak, but there are

The ammonite (**Above**), with its coiled shell, was the dominant animal of the Mesozoic seas 150 million years ago. It is very similar in appearance to the modern nautilus (**Far Left**). In fact, the nautilus was around long before the ammonites but the shells of the early nautiloids were straight cones up to five metres (16.5 feet) in length. It was only later that the shell became coiled.

Above Left: Mussels are bivalves with two hinged shells that can be clamped together by strong muscles.

Shrimps (**Top**) and prawns are laterally flattened crustaceans whereas the crab (**Above**) is dorso-ventrally flattened.

Top: The octopus (**Left**) and the cuttlefish (**Right**) are both cephalopods. They have tentacles covered in suckers with which they catch their prey. They can move quickly through the water, relying on their highly efficient water-jet propulsion systems.

Above, Left and Right: All the members of the echinoderm family have a spiny skin. Although they all look very different, their bodies are all based on a radial plan. Sea lilies, or crinoids, are stalked echinoderms with branched arms covered in cilia for filter-feeding.

More members of the echinoderm family. The sea cucumber (**Top**) has a tough leathery skin. The starfish or sun star typically has five arms, but some (**Top Right**) have as many as 14. The radial pattern of the sea urchin is more easily seen on the skeleton (**Above**) than on the living animal (**Above Right**).

NUDIBRANCHS

There are two types of nudibranch. The aeolids (**This Page, Opposite and Page 34**) feed on cnidarians such as corals, sea anemones and hydroids. These nudibranchs that have sting cells for protection and long tubular projections on their backs called cerata. At the tip of each of the latter is a sac in which the supply of stinging cells is stored, the sting cells having been taken from the bodies of the cnidarians. In fact, some aeolids steal the zooxanthellae found in coral polyps and nurture them in their own tissues in order to gain more food.

The dorids feed on sponges (**Page 35**). Sponges contain toxic compounds and when these are eaten, the nudibranchs absorb and concentrate the toxins in their own tissues. Each sponge may have many different toxins in their body and the nudibranch has the ability to concentrate the four or five most toxic compounds. This way, the nudibranch becomes poisonous.

The presence of such toxic substances in the sponge is also for protection. Approximately 99.9 per cent of the sponge larvae are eaten, but once they become adults they are protected by their toxins. This works against most reef animals except, of course, the nudibranch. The pharmaceutical industry is currently studying the sponge toxins as part of its hunt for new antibiotics and anti-cancer drugs.

hundreds of them and, together, they are capable of moving the starfish over the floor. They can even exert enough pressure to force apart the shells of a bivalve.

One of the most insignificant organisms found on the sea shore is a shapeless lump known as a sea squirt. This jelly-like creature has two connected openings and is called a squirt because it squirts a jet of water up the leg of anyone who accidentally steps on it. Its is important in evolutionary history because it is one of the earliest ancestors of the vertebrates—the group to which fish, amphibians, reptiles, birds, mammals and, of course, humans belong. The sea squirt larva looks just like a miniature tadpole: it can just about swim through the water and is supported by a flexible rod running the length of its body. It is carried away from the adults on the currents and will settle down on another rock and grow into maturity. This tadpole-shaped larva, with its primitive spine, is related to the very earliest fish.

Evolution in the Seas

The fish that most resemble their earliest ancestors are lampreys and hag-fish. These species can barely swim and, instead of jaws, they have a sort of sucker. Their heads end in a flat disc, at the centre of which is a tongue that is covered in sharp spines. They can clamp this disc onto the bodies of other fish and chew through flesh using the tongue—slowly eating the prey alive. Fish like this first developed more than 500 million years ago and over the next 100 million years or so developed into a more modern looking version with gills and jaws. Their teeth were formed from specialised scales in the skin, while flaps of skin also grew into fins, giving more control over movement through water. This ancestral type of fish gave rise to the two modern groups that we find in the oceans today—cartilaginous fish such as sharks and rays, and bony fish such as cod, herring and salmon.

Cartilage is a far more flexible material than bone; in fact, it is the material that makes up our own noses and ear flaps. Fish skeletons have to be both lightweight and elastic. The shark, for example, has a long tail that thrashes through the water from side to side, propelling it forwards. The body design of the ray is different to that of the shark in that it is much flatter, with pectoral fins like large sails and a tail that is reduced to just a whip.

Bony fish, on the other hand, have retained a hard, inflexible bony skeleton and, unlike sharks, they have a buoyancy aid—a swim bladder. By altering the amount of air in the swim bladder, the fish can move up or down in the water. Their body shape is generally more streamlined than the shark, and their scales are covered in slippery mucus. These factors allow such fish to slip easily through the water. The fastest of the bony fish are tuna, marlin and bonito, all of which have a long tapered body and are powered through the water by rows of muscles.

Fish have evolved into a highly diverse and successful group of animals—the number of different species of fish is approximately the same as all other vertebrate groups combined. They can be found in all watery habitats, from tiny pools of water in the desert, to the icy waters of the poles and the deep cold world of the ocean depths. A few fish can even survive on land for short periods of time. Mudskippers, which are found in mangroves and swampy coastal areas, have evolved stiffened pectoral fins that allow them to flop across the mud from pool to pool and they hold water

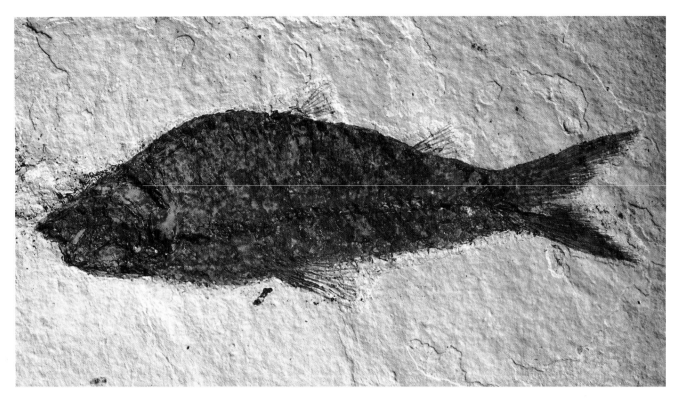

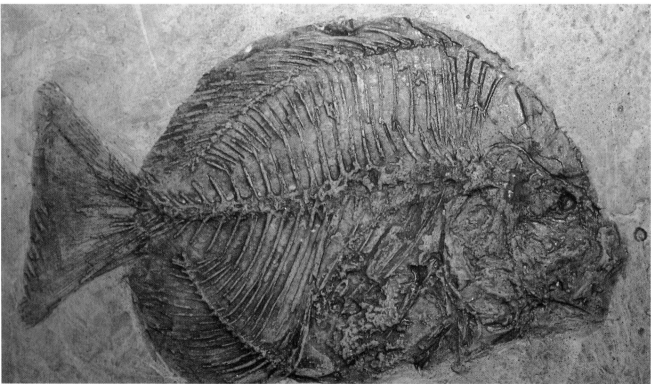

in their mouth to provide oxygen. However, they can do this for only for a short period of time before they have to return to the water. From this example it is not too difficult then to imagine how some fish developed lungs and stiffer limbs, eventually evolving into amphibians, the group of animals that includes toads, frogs and salamanders.

All amphibians have a moist skin that restricts them to damp habitats and they must return to water to breed. They cannot survive in salt water but some of their descendants, the reptiles, can. Most reptiles are also terrestrial, but there is a group of marine reptiles found on the Galapagos Islands off the coast of Ecuador. There is little vegetation here and during the day sunlight heats the

Left, Above and Below: The earliest fish fossils date back some 500 million years. These fossil fish are much more recent, being only about 50 million years old.

The sting ray (**Top**) and the sand tiger shark (**Above**) are both examples of cartilaginous fish.

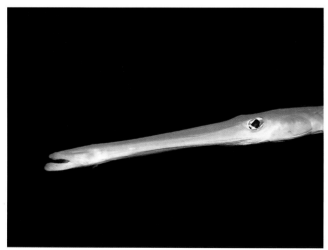

The range of shape and colour in fish is very diverse, ranging from the elongated cornet fish (**Above**), the irregularly shaped sea horse (**Top**), the laterally flattened bannerfish (**Above Far Right**), the more common rounded shape of the ballen wrasse (**Far Right**) and the cardinal fish (**Right**) with its exaggerated fins.

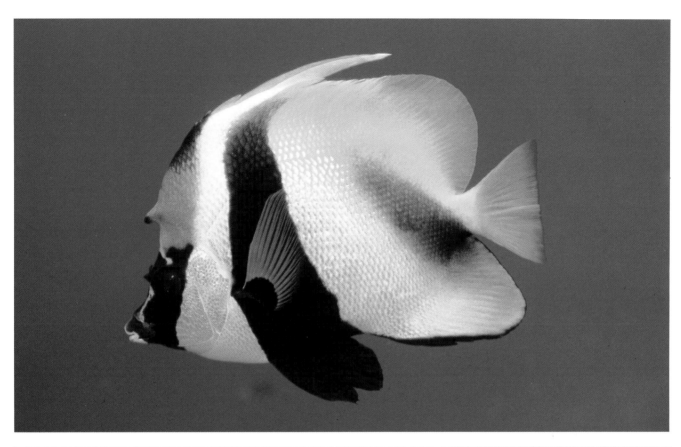

41

dark volcanic rock on which the marine iguanas make their home. For much of the day, these lizards lie on the rocks so that their bodies warm up. In order to keep a constant body temperature they shuffle around, first lying at right angles to the sun and then head on. If they get too hot, they seek out what little shade there is between the rocks. The waters around the islands are cold, so the iguanas only go into the sea when it is so hot they cannot cool off any other way. During their cooling swim they feast on seaweeds and then they clamber back onto the rocks to warm up again.

There are other marine reptiles. Saltwater crocodiles, for example, live in the brackish waters of swamps and estuaries. But the most widespread marine reptile is the turtle. Just like the tortoise, the turtle has an armour-plated body made of horny plates that completely encase its body and make it very heavy. On land, turtles and tortoises are slow moving but, once in the sea, turtles can move with ease. They have one problem—their leathery eggs are unable to survive in water, so the female turtle has to return to land in order to lay them.

Millions of years ago, a group of chicken-sized reptiles developed feathers on their fore limbs and gained the ability to fly. Today, birds rule the skies and though they are tied to land, for they have to have somewhere to build their nests and lay their eggs, many sea birds rely on the oceans to provide them with their food. A diversity of birds can be found in coastal areas, especially estuaries, where they feed on the wealth of life hiding in the mud. In the Antarctic, birds are the dominant life form. This is the home of the penguin, a bird that has lost the power of flight but can swim as well as many fish.

Mammals are the most highly evolved group of animals and evolved from very early reptiles—the link between the duck-billed platypus, which lays leathery eggs, and reptiles is evidence of this. Mammals, just like birds, are endothermic. This means they can control their body temperature to keep it within a narrow range, making use of heat generated internally. Their bodies are covered with hair and, as the name suggests, the female mammal has mammary glands that produce milk to feed her young. Most mammals live on land, but three groups—the cetaceans (whales and dolphins), the sirenians (manatees and dugongs) and the pinnipeds (seals and sea lions)—have returned to the sea and their bodies have become adapted to life in seawater.

Top: A Galapagos Island marine iguana sits on the rocks until it is too hot, when it plunges into the cold water to cool off.

Above: The ichthyosaur is a fossil reptile that swam in the Jurassic seas 170 million years ago. At that time dinosaurs ruled the land.

The turtle (**Right**) and the salt-water crocodile (**Above Right**) may look very different, but both lay eggs on land.

Left: The wandering albatross spends much of its life at sea, gliding over the surface of the oceans in search of food. It only returns to land to mate and raise a chick.

Right: Dolphins and porpoises are fast moving animals, cutting through the water at speeds of more than 40 kilometers per hour (25 miles per hour). They find their way using echo location, or sonar, a similar system to that used by bats flying at night. The dolphin produces a stream of high-pitched clicks that is reflected by objects in the water in front. A dolphin's sonar is so sensitive that it cannot only tell where an object is, but also what it is.

Below: One of the most important fossils in existence is that of Archaeopteryx—the first known animal to have feathers. It is thought to be one of the links between reptiles and birds.

EYES

Who would think that eyes could be so fascinating and so varied? At least some underwater photographers think so and have made this colourful study of some of the eyes found on animals that live in the ocean. In this watery world, vision is different from how we see on land; water is a thicker medium than air, making the image blurred and distorted. When we swim underwater we usually have the benefit of goggles to give us a layer of air before our eyes. It is interesting to ponder on what these sea creatures are actually seeing. Do they see in colour as we do? Are objects magnified? Do they see a mosaic picture like the compound eyes of insects? Will we ever know exactly what they see?

Above: This is the stalked compound eye of the red coral crab.

Below Left: The crayfish is a crustacean—like the crab—and its stalked compound eye is clearly visible here.

Right: Cuttlefish eyes are fascinating as they are very similar to those of vertebrates yet the cuttlefish is an invertebrate mollusc. The eyes have followed a parallel course of evolution but the eye is derived from the skin rather than the brain.

Overleaf: This eye of the honeycomb moray eel is very like that of the parrot fish. Moray eels need to have good vision as they hide among the coral with only their head visible as they look out for prey.

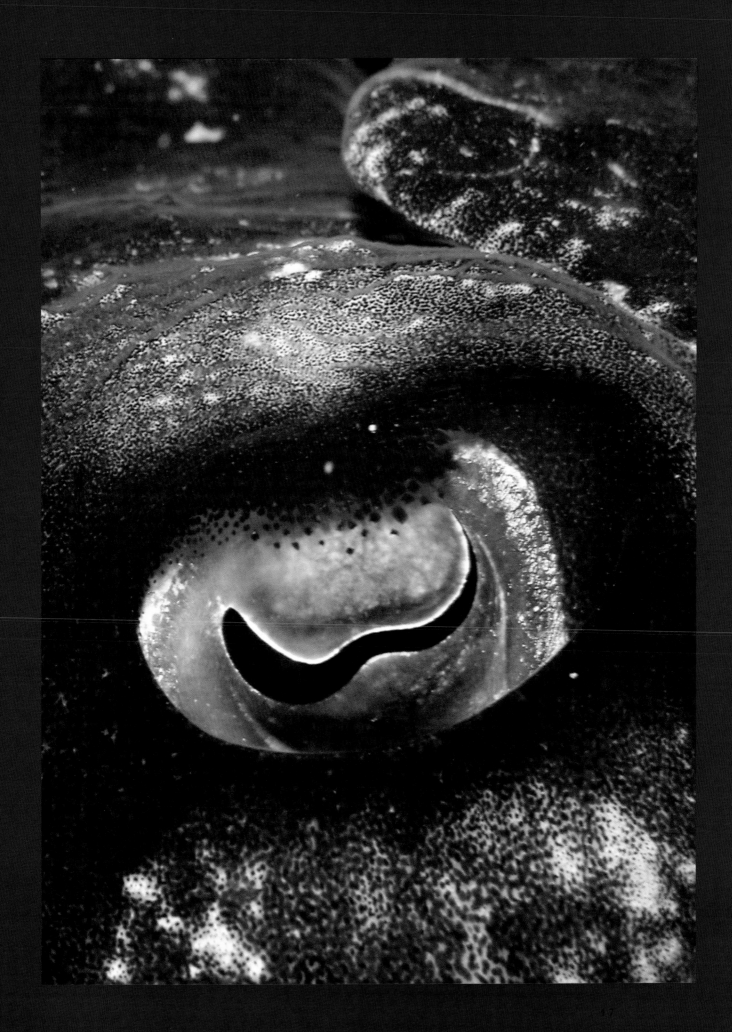

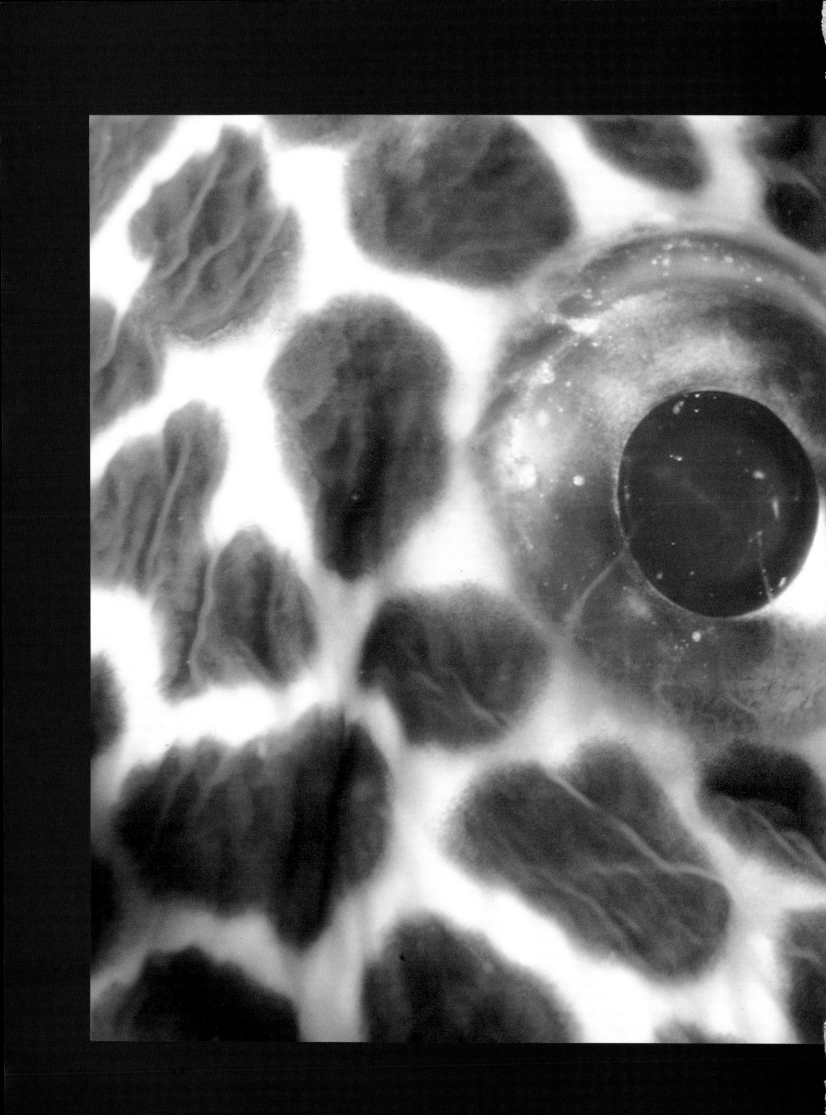

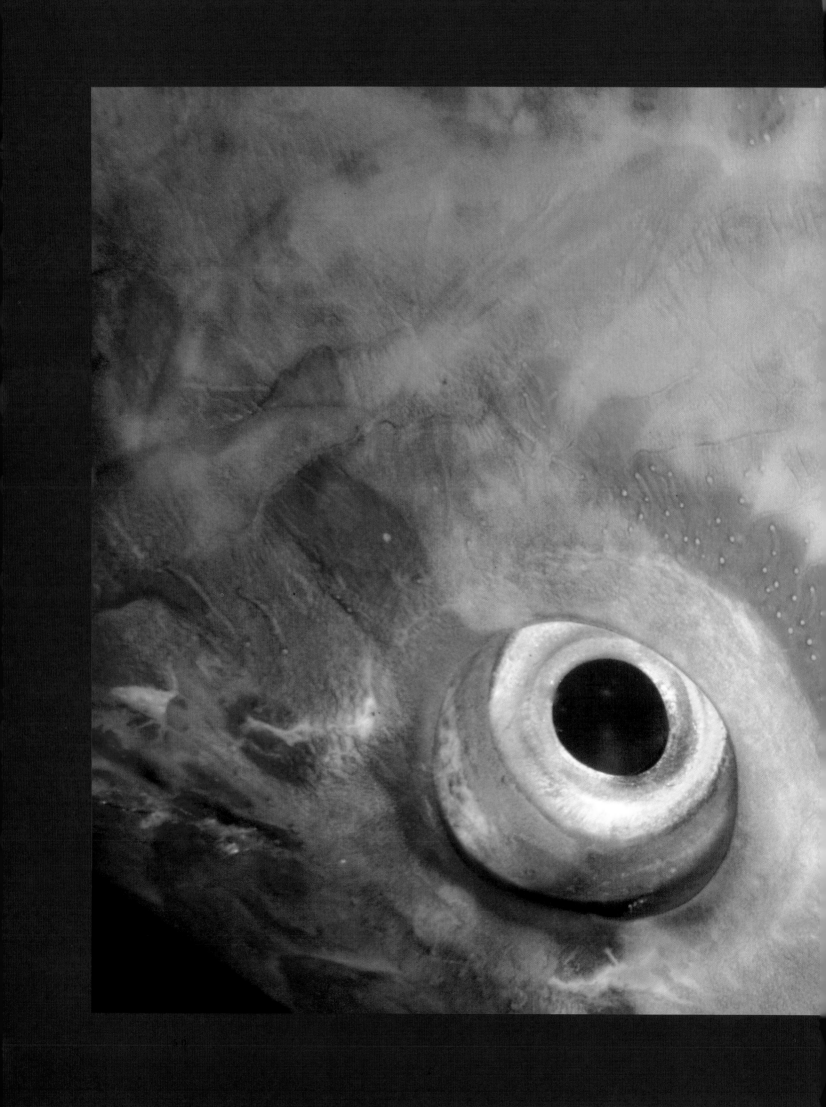

The eyes of the parrot fish are basically similar to the eyes of other vertebrates like ourselves. However, the transparent lens is more nearly spherical than those of land animals, which gives a wider angle of vision.

Although these three groups have different origins they have all overcome the difficulties of living in water in the same ways. One major problem is keeping warm, which is why marine mammals tend to be large with a small surface area relative to their volume, an attribute that cuts down on heat loss. Whales, for example, are huge animals that have been able to grow so large because of the buoyancy of the water. The blue whale is the largest living animal on earth, reaching over 35 metres (115 feet) in length and weighing a massive 130 tonnes—the combined weight of about 25 elephants. Heat loss can be further reduced by having a thick layer of fat, or blubber, that acts as insulation. In whales, the blubber can be a metre (3.3 feet) thick. Their bodies have a streamlined shape to reduce drag through the water and their limbs are much reduced in size. Whales and dolphins have lost their hind limbs completely, while pinniped hind legs are much reduced in size. Marine mammals are constrained by the fact that they still have lungs and have to return to the surface to breathe quite regularly—although some can hold their breath for up to two hours by closing off their air passages and surviving on low levels of oxygen using anaerobic respiration in their muscles.

Above: The only time whales are spotted is when they come to the surface to breathe, as this humpback whale is doing.

Top: Seals and sea lions are mammals that live in coastal waters, but they have to return to land to breed. They are well adapted to living in water, with a thick water-repellent fur that insulates the body and gives it a streamlined shape. They can only waddle on land, but are graceful swimmers and deadly hunters underwater.

DIVERS VERSUS THE SEAL

A human preparing to dive automatically takes a deep breath, and holds it underwater. The air is expelled slowly, and this limits time underwater to just a few minutes. Somebody undertaking a deep dive runs the risk of developing the bends as they return to the surface. If the diver rises too quickly, bubbles of nitrogen gas form in blood vessels and block the flow of blood, causing injury and even death. A seal tackles the problem of diving in a very different way. At the beginning of its dive, a seal will actually collapse its lungs. The lack of air in the lungs not only helps to reduce buoyancy so they can dive deeper, but it also prevents the nitrogen forming bubbles. The seal also cuts off blood flow to unessential parts of the body, such as the digestive system, and directs as much blood as possible to the muscles and brain. Their heart rate slows, as does their metabolism, so their body is effectively just "ticking over" very slowly.

Above: A snorkeller dives down to a coral reef. The human lung cannot collapse, so humans run the risk of developing the bends if they come to the surface too quickly after a prolonged dive.

Ocean Environments

The marine ecosystem can be divided into two main habitats. The water itself is called the pelagic habitat, while the bottom is known as the benthic habitat. Each is colonised by a different community of plants and animals.

The pelagic habitat is divided into three parts according to the level of light. The euphotic zone is the surface layer where there is plenty of light; the bathyal or twilight zone has limited light; while the deep areas of the abyssal zone are almost completely dark.

Light does not pass easily through water because it is absorbed by suspended plankton. The depth to which the light can penetrate varies. In productive seas and oceans, there is an abundance of plankton that prevents the light from penetrating more than 100 metres (330 feet) or so whereas light can penetrate several hundred metres in the clear waters of the tropics, for they are much less productive and almost devoid of planktonic life. The depth of light penetration sets the limits to the distribution of plants. Marine plants all belong to the group called algae. Some algae

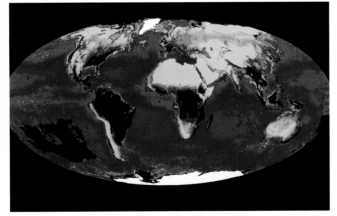

Above Left and Above: The shallow green waters around these tropical islands are more productive than the deeper ocean.

Left: This false-colour image of the earth shows the productivity of the different ecosystems. Coastal regions, shown in red, are the most productive, while the darkest blue regions of the open ocean are the least productive.

are single-celled organisms and form part of the plankton, while others are seaweeds and are found in the coastal zones. Like all plants, they need light for photosynthesis, the process by which they make food. The algae are at the bottom of the oceanic food chains so their productivity affects the overall abundance of marine life.

The highest productivity occurs in areas such as the shallow coastal waters over the continental shelf. Here, tides and storms keep the water well mixed and bring up nutrients from the floor where it mixes with runoff from rivers. In temperate regions, there are two peaks of productivity—spring and autumn. The low light levels and temperatures of winter restrict productivity and during

the summer, the surface layer of water warms up and sits over a cooler layer below. This creates a thermocline and the two layers do not mix. The nutrients in the warm layer become depleted and it is only in the autumn, when the thermocline breaks down, that production can resume. In the Southern Ocean around Antarctica, the water is cold and rich in oxygen and nutrients. There is no thermocline because water is mixed vigorously by strong winds, so there is extremely high productivity during the short Antarctic summer. The vast areas of deep oceans in the tropics have low levels of nutrients and this limits the plant productivity. However, although the production may be lower, it is continuous.

Above: The rough seas of the Southern Ocean are amongst the most productive in the world.

Left: It takes very specialised groups of animals to survive in the hostile environment of the deep sea floor, such as this pale coloured spider crab that is only found near thermal vents.

Right: The tides have eroded this rocky shore in South Africa, creating sandy bays and rocky headlands. When the tide goes out, small pools are trapped between the rocks (**Overleaf**).

PLANKTON AND NEKTON

Animals of the open ocean fall into two groups; plankton and nekton. Planktonic animals are drifters that are carried by the current, whereas nekton are active swimmers. Nekton include squid, fish and other marine mammals like whales. However, even the active swimmers have found ways of floating in the water, which are useful because it reduces the amount of energy required to move. Nektonic animals are larger than planktonic animals and have sense organs. In general, they are nearly all carnivores, feeding on the plankton or benthic animals.

Fish, such as this black jack (**Above**), are active swimmers so they are in the nekton group. Jellyfish (**Below**) are the most common animals of the open ocean, making up much of the plankton.

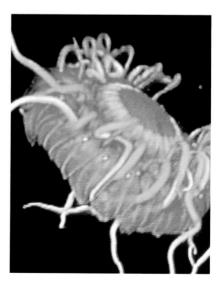

Above: A large expanse of bare sand at low tide. All the animal life on this shore is buried in the sand waiting for the tide to return.

Beneath the relatively brightly lit layer of the euphotic zone lies the twilight bathyal zone, generally reaching from 200 metres (650 feet) to 2,000 metres (6,500 feet). The deeper the water, the higher the pressure and because there is very little light, there are no plants. Consequently, the animals that live in the bathyal zone are adapted to both high pressure and a limited supply of food. Food sources, such as plankton, waste materials and dead animals, decrease with depth, so the animals living in this zone are highly efficient predators able to survive on an infrequent food supply.

The abyssal is the part of the ocean that is found more than 2,000 metres (6,500 feet) below the surface. It is cold, dark, under great pressure, and has even less food. The open water at this depth is almost devoid of life but the very few creatures that are present are highly efficient predators and have lost almost all their inessential organs. They tend to have small, grey-coloured bodies and are often blind. However, they have highly developed senses of touch and smell, and have specialised guts to obtain the maximum nutrition from the paltry supply of food.

The benthic zone is the actual seabed and ranges from above the high tide mark on the shore to the deepest parts of the ocean. Plants are only found in shallow waters because they are limited by the amount of light penetrating the water. The benthic environment varies from sand, coral and solid rock to ooze and mud and each type of substrate supports a different community of animals.

The uppermost environment is the inter-tidal zone, that region of the coast regularly exposed to the air by the tides. The plants and animals found here are adapted to the extreme conditions created by the tides; one day covered by the water and subjected to wave action, the next exposed to the drying effects of the sun and wind. Rocky shores are dominated by seaweeds that attach themselves to the rocks. The type of seaweed depends on its position on the

Above: Each time a wave crashes onto the shingle beach water churns the pebbles and makes it almost impossible for animals to survive.

Left: There is much greater biodiversity on a coral reef than in the open ocean.

shore, with those higher up being more tolerant of desiccation than those lower down. The animals living in and amongst the rocks are mainly sedentary and include barnacles, mussels, limpets and fan worms. These creatures either filter the water for suspended particles or graze on the algae, although there are also a number of predators such as crabs, starfish and dog whelks that thrive in this environment. Sandy shores are characterised by seemingly barren expanses of sand and pebbles, but there is a surprising diversity of life buried beneath the sand.

The abundance of animal life depends on the amount of available food. The seabed in the shallow waters has the highest productivity and, in general, the amount of living material or biomass declines with depth. The deep seabeds are populated by very few animals, with the exception of the very diverse and specialised group of organisms found around the hot, mineral rich vents in volcanic areas. Thousands of metres below the surface, a unique and truly fascinating group of animals have recently been discovered. These creatures use the chemicals in the water as a source of food, and every expedition to the sea floor reveals more about their precarious existence.

At the edges of the ocean there are wetland habitats such as estuaries, mangrove swamps and salt marshes. As tides rise and fall and rivers flow, the salinity of the water in such places ranges from near fresh to very salty. There are copious amounts of mud, organic debris and nutrients which are colonised by a disparate and highly adapted range of animals that can tolerate the widely different conditions. The changing salinity, the shifting mud and the varying temperature all make it a very challenging environment in which to survive, but the vast amount of food that is available means that productivity is high. These areas of the coastal ecosystem are also very important fish nurseries, for the young fish fry can get plenty of food to fuel their growth. Regrettably, however, these coastal areas are where people have had a considerable impact and have caused a lot of pollution, often with devastating consequences.

In the chapters that follow, you can read about the plants and animals that live in the widely differing habitats of marine ecosystems and learn how they have become adapted to their environment. This voyage of discovery starts at the very edge of the ocean, looking at the animals and plants that live along our shorelines. Then you will

be taken into the ocean itself and experience the wonders of this watery world, the variety of plant and animal life and the beauty of the coral reef. Diving deeper still, you can share the scientific research which has led to the discovery of a vast new ecosystem at the bottom of the deepest oceans, with bizarre forms of life that seem to come straight from science fiction. Finally, you can consider the impact of humans upon our beautiful watery planet, and become aware of the conservation measures necessary to rectify much of our devastation of the oceans; and ultimately our own lives.

Above: Mangrove trees are specially adapted to survive on the mud of this tropical swamp.

Above Left: The vast expanse of mud exposed at low tide is full of worms and molluscs that provide a ready source of food for wading birds.

Left: The sun rises over a mist-shrouded Lafoten Island in the Arctic Circle.

Where the Ocean Meets the Land

When we think of ocean life there is a temptation to think big and deep. In the mind's eye we see aquamarine, vivid coral gardens, myriad rainbow fish and dolphins somersaulting through the foam. But let us start at the place where the ocean meets the land. Nowhere is quite like it. Every time a wave comes crashing up a beach, anywhere in the world, a drop of ocean touches the sand and there is life in that spray. The world's shorelines stretch for many thousands of kilometres around every island and continent and are as diverse as any inland habitat. From sandy beaches and dunes, shallow lagoons, sticky mudflats and tropical mangrove swamps to dramatic rocky coastlines with towering cliffs and the hidden delights of their rock, or tide, pools each of these environments supports a great wealth of ocean life. Animals and plants must be highly adapted to cope in this fragile world of the tidal zone as it is very different from life in the open ocean but when we start exploring we find that a rich variety of plants and animals successfully colonise this habitat.

Standing by the edge of the sea on a sandy beach, there is a profound sense of antiquity for anyone who is aware of the great passage of time that such places take to form. The sand on most seashores has been derived from the weathering and decay of rocks, carried to the sea by

Continued on page 73

Right: This dramatic rocky coastline in Cornwall, England, is washed day and night by waves. Along the seashore water is always moving, either gently along the beaches or crashing against cliffs and rocks.

THE TIDAL CYCLE

Twice every 24 hours and 50 minutes the water level in the oceans rises and falls as tides. The tides are the response of the waters of the ocean to the gravitational pull of the moon and the more distant sun. This gravitational pull causes the surface of the water, on the side of the earth facing the moon or sun, to be pulled away from the solid mass of the earth, causing a bulge. Another bulge occurs on the far side of the earth in the opposite direction, and these bulges move around the planet as the earth rotates, forming high tides. Spring tides are the highest and lowest tides and occur when the earth, moon and sun are in a straight line. At that time, the gravitational pulls of the sun and moon are combined. Neap tides are found when the difference between high and low tide is at its lowest. They occur when the moon, sun and earth form a right angle and therefore the gravitational pulls of the moon and sun partly cancel each other out.

Above, Left and Right: Most of the shore along this rocky coast in Wales is covered by the sea at high tide. At low tide much of the shoreline is exposed. The tidal range is the difference in height between the high and low tide. The average tidal range is between two and three metres (6.5 and ten feet). However, enclosed seas like the Mediterranean are almost tideless.

Above: The classic image of a deserted sandy beach on a tropical island in the Maldives.

Overleaf: A sandy beach can be a treasure-trove of fascinating empty shells for the collector.

Top: A harbour seal hauls itself up on a sandy beach in California.

Above: An expanse of sandy shoreline showing very little sign of life.

Above Right: A ghost crab ventures from its burrow on a sandy beach in the Indian Ocean.

rain and rivers over countless thousands of years. Most beach sand consists of quartz—the most abundant of all minerals quartz is found in almost every type of rock—although in some parts of the world the sands are made up of the remains of coral or fragments of the chalky shells of sea animals. Sand is always on the move as waves wash over its surface so no plants can anchor themselves on the shifting substrate. However, a wealth of animal life can be found, though it is mainly revealed by digging.

A thin film of water is held around each grain of sand. This prevents the sand grains from rubbing against each other and permits animals to live within the sand after the level of the sea has fallen below it. This tiny world of the sand grain is the home of minute organisms. There are single-celled animals and plants, water mites, insects and the larvae of tiny worms, which live, breathe, feed, reproduce, swim and die in a world in which each minute droplet of water separating one grain of sand from another is like a vast ocean. It should be noted, however, that not all sands contain this "interstitial fauna"—shell or coral sands seem to create unfavourably alkaline conditions in the water surrounding the grains.

Sandy beaches not only look uninhabited, but uninhabitable. However the discerning eye can observe clues to the population of this environment. Slight tracks might be seen, a tiny movement disturbing the upper layers or barely protruding tubes leading to hidden burrows. The animals are beneath the surface of the sand, hiding from sun, wind and predators. As the tide rises and covers the beach they come out to feed on nutrients carried in by the tide.

The upper shore, which is continuously uncovered is the most difficult zone for occupation by marine plants and animals and has the least species. Sand on the upper shore is dried by the sun at low tide and some may be blown by the wind and build up beyond the strand line, eventually forming sand dunes. The drifting sand is a difficult place for plants to grow, but a few species such as sea couch grass and marram grass are able to establish themselves on the dunes and they help to prevent the sand from blowing away. These grasses stabilise the sand allowing other sand inhabiting plants such as sea holly to become secure. Once these plants have stabilised the habitat, birds can begin to nest in the dunes and small mammals, reptiles and insects can make a life there.

In contrast, the middle shore, which experiences twice daily submergence, has an abundance of wildlife. Probably the most numerous animals on sandy beaches are the bivalve molluscs. For safety these creatures burrow vertically into the sand with great speed but poke two tubes into the clear water above, with which they suck a current of water through a filter between their shells. Certain carnivorous snails also burrow in sand, together with great numbers of crustaceans, chiefly of the "hopper" type. This burrowing habit provides shelter, escape from predators, avoidance of desiccation, conservation of water and provision of food on the dry, shifting sandy shore. There are also burrowing starfish, sand eels and other sand burrowing fishes, while amongst the scavengers and carnivores can also be found weever fish and the common goby.

The incoming water at high tide brings a fresh supply of food and oxygen to the middle shore. The crabs, flatfish and common goby fish that have come inshore join the crustaceans, molluscs and worms to search the surface for food. During low tide, however, the exposed sand and its inhabitants provide a feast for wading sea birds, which are the top predators in this marine food chain.

Above: Wood, seaweed and other jetsam are washed up on the Pacific coast of the USA. This debris forms the strandline, the habitat of scavenging sandhoppers—marine animals that have become almost terrestrial—together with many land-living insects that feed on the organic debris.

Left: Oyster catchers feeding at the exposed low tide line.

Far Left: The holes in the sand were made by feeding turnstones as they hunted for bivalves during low tide.

Whilst sandy shores are composed of small grains of sand, on muddy shores the smallest particles of all are found—silt and clay. As there is almost no slope on muddy shores they are usually called mudflats. The lack of water movement in such places allows the mud to build up and it never dries out completely, even at the surface, when the tide retreats and so it forms a highly stable medium in which animals are able to make permanent burrows.

The organic matter in the mud provides food for many burrowing animals such as the lugworm. However, organic matter can be a disadvantage as it removes oxygen, normally present in water contained in sand, so that life becomes difficult at any depth below the surface. Burrowing animals must continually draw water into their burrows but the fine particles of mud tend to clog delicate gills and tentacles used for breathing and feeding and consequently there are fewer animals that are capable of living in mud.

All the rivers of the world are heavy with sediment by the time that they reach their estuaries. Here, the sediment is deposited to form a vast low-lying area of mudflats and sandy banks. Estuarine mud has a fineness, a stickiness and a smelliness all of its own, as the gases produced by the decomposition of the organic debris stay trapped within it. The conditions range from low to high salinity, with a corresponding gradation of plant and animal life. When the tide goes out, especially when rivers are swollen with rain, fresh water predominates but when the tide comes in the water in the estuary becomes as salty as the sea. Organisms that live in this habitat must therefore be able to withstand a great range of chemical and physical conditions. However, their great reward is that food is delivered to the estuary daily from both sea and land. Estuarine water is potentially more nutritious than almost anywhere else on the seashore and this is why the few species that can survive in these conditions flourish in immense numbers. For example, in just one square metre (eleven square feet) of mud there may be as many as 250,000 sludge worms or 42,000 spire worms!

Above: The marine lugworm, Arenicola, is about 40 centimetres (16 inches) long and as thick as a pencil. It digs a U-shaped tube lined with mucus. Gripping the sides of the tube with its bristles, it moves up and down at the bottom like a piston and sucks water through the sand plug. Particles in the water are trapped in the sandy mud and the worm digests the edible bits, egesting the rest as the familiar worm casts.

Left: Marram grass is a vigorous plant that produces extensive underground stems called rhizomes. Its dense growth slows down the wind speed over the surface of the sand and stabilises the sand dune. Here it is found growing in a sand dune in southern Spain.

Overleaf: This aerial view of an estuary in northern Queensland, Australia, shows the mosaic of habitats typical of estuaries, such as channels, mangroves, sandbars, inter-tidal flats and beach ridges.

When the tide retreats all the animals living here must stop feeding and find ways to prevent desiccation. Spiral shells seal the entrance to their shells, cockles clamp the two halves of their shell tightly together and lugworms retreat into their burrows. Unfortunately desiccation is not the only problem facing these mud dwellers, for huge flocks of hungry birds descend upon the estuaries at low tide to feed.

As the rivers bring down more and more sediment, so the mud-flats slowly rise and a film of green algae begins to form on them, binding the mud particles together. Once this happens other plants can get a root hold and the mud banks begin to grow at an amazing speed as the roots and stems of the plants trap even more mud. Eventually the surface of the mud becomes sufficiently far above the water to be beyond the reach of all but the highest tides. Around European shores, the pioneering reclamation of the land is carried out by a small plant called glasswort.

Above: A banded stilt feeding in the muddy waters of the estuary. In a day's feeding a small wader can consume several hundred bivalve molluscs or thousands of crustaceans.

Above Right: Glasswort is a halophytic plant with scale-like leaves and swollen translucent stems. It has thus adapted well to the extreme salty conditions of the salt marsh.

Right: The salt resistant vegetation of salt marshes is not as productive as that of the tropical mangroves, but forms a vital part of estuary environments in northern latitudes.

Salt marshes form in the more sheltered regions of the coast, where the action of the tide is minimal. Green seaweeds are often the first plants to colonise the mud; their fronds trap muddy sand and once this process has started glasswort, cord grass and eel grass can then become established. On the higher parts of this now well-formed marsh other flowering plants will appear, such as sea lavender and sea aster.

Left: Where there is a great seasonal variation in the flow of a river, an estuary mouth may close during the dry season and wave action may quickly throw up a sand bar. The sand bar reduces wave action and creates an enclosed lagoon.

Overleaf: These mangrove trees are growing in tropical wetlands in southeast Queensland, Australia.

The tropical equivalent of the salt marsh is the mangrove swamp. These are places of strange beauty, particularly at low tide when tangles of arching roots and rows of spikes emerge from the mud. There are many different types of mangrove trees and shrubs, each adapted to a particular salinity or water depth, to sand or muddy substrates but keeping a foot-hold on the shifting mud is a major problem for all of them. The warm mud just below the surface is low in oxygen and very acidic so the trees cannot send down deep roots. Instead, the mangroves have roots that resemble floating platforms that sit like a raft on top of the mud. Rather than taking oxygen from the mud the mangroves either draw it directly from the air through spongy patches on their bark or through aerial roots that rise vertically out of the mud. At low tide water drips from the roots, while clicks and pops sound in the heavy tropical air as molluscs and crustaceans snap their claws and close their shells. Mudskippers plop from pool to pool and an army of animals comes out of hiding to gather the food left behind by the receding tide. As the waters flood back, the tangle of roots disappears and the mangrove forest is transformed.

Pebbles are much larger than sand and are composed of smooth rounded pieces of rock ranging from about half a centimetre (one fifth of an inch) to 12 centimetres (five inches) across. A beach made of pebbles is known as a shingle beach and such beaches are characterised by a very steep slope of about 12 degrees due to the waves rolling the pebbles over each other as they rush up and then fall back over the beach. Shingle beaches are the most unstable of

Above: The pebbles of this shingle beach in Cornwall, England, are continually moved by the waves.

Above Left: This mangrove land crab is standing by its hole in Costa Rica.

Right: Dramatic limestone rock formations are found along the rocky shoreline of southern Thailand.

shores as water cannot be retained in the spaces between the stones when the tide falls because the pebbles are so big. Everything on the surface is crushed by the constantly rolling pebbles so it is incredibly difficult for plants and animals to live on or within a shingle beach.

Although vegetation cannot establish itself on the shoreline, where the sea moves and washes the shingle, a little further up plants that tolerate salt spray appear. Higher up the beach still, the stones are more stable and soil begins to accumulate between them so a wider variety of plants can become established. Several species of birds may also be found breeding on the shingle beach, laying their eggs among the pebbles and fishing just offshore.

At first sight, rocky shorelines seem as inhospitable as the shingle beach but in reality, conditions are very different. Rocky coastlines are by far the most variable in character of all the shores depending on the type of rock there, hard or soft, igneous or sedimentary, and if the latter, on the nature and slope of the strata. Undoubtedly the most attractive of all shores are those where rock or tide pools occur.

Probably the first thing that a visitor to a rocky shore notices is the presence of distinct areas or bands running along it, appearing as light or dark strips. The shore is actually the region between the extremes of tidal movement and it may be divided into a series of horizontal zones, according to the extent to which each is covered

Above: Bladder wrack and knotted wrack carpeting the rocks during summer in Dorset, England.

Right: This rocky coast shows clear zonation with the orange lichen forming the uppermost zone with a band of black lichen beneath it. The seaweeds are found lower down the shore where they are covered by the tides.

and uncovered during the full cycle of tidal movements. The zonation that occurs on a particular shore is a reflection of the interaction between the terrestrial and marine animals and plants that live there. Two of the key factors are the tolerance of animals and plants to submergence and desiccation, and the amount of wave action they can withstand. Those organisms that cannot tolerate exposure for long periods tend to be found low down the shore, while those that can withstand long periods of exposure, and the consequent drying out, are found much higher up. Zonation, therefore, really involves those animals and plants that are fixed and cannot move. However, when the tide has covered a shore, various animals not normally seen on the exposed part, particularly fish, move in to feed. Similarly when the shore is exposed after the tide has receded, birds and rats may visit looking for food.

The upper shore is the area above the average high tide level and it is continuously uncovered except when the tides exceed average range. It is therefore the most difficult zone for marine plants and animals to occupy, but although it is the most sparsely populated in variety of species, it is not necessarily in numbers of individuals.

On the other hand, the middle shore represents typical shore conditions and the animals and plants that occur here in great abundance—both of species and individuals—are the inhabitants that most characterise the shore. This is an extensive region bounded above by the average high tide level and below by the average low tide level. Therefore it experiences twice daily submergence under the sea and then exposure to the air.

Finally, the lower shore is only uncovered when the tides exceed average range, so it the easiest region for marine animals to colonise, and seaweeds have less danger of drying out. It is a narrow region extending from the average low tide level down to the limit of the shore at the extreme low water level of spring tides. Onto it extend animals and plants from the area below tidal levels and it is thus difficult to make a clear cut definition of where the land ends and the ocean begins.

There is always competition for food and space and in order to compete successfully for these two commodities a species must outperform others in growth rate, adaptive behaviour and feeding efficiency. Sedentary animals need to occupy the right site if they are to survive, and their location is often selected while they are in their younger planktonic floating stages. The distribution of mobile animals on the shore, such as crabs, depends upon their ability to remove themselves from distressful environmental factors. Crevices, for example, may offer protection from the pounding waves for such creatures. The zoning of plants reflects their ability to withstand desiccation when uncovered, buffeting from waves and the material thrown up by waves and the grazing and browsing activities of shore animals.

Above: A red-horned sea-star in the shallow water at the low tide mark in Northern Territory, Australia.

Above Left: The snakelocks anemone is found in a rock pool on the lower shore where it will never be exposed by the tide.

In the hollows among the rocks, where water remains even when the tide is out, rock or tide pools are formed. The organisms which inhabit them include many of those already encountered on exposed rocks. Seaweeds line their walls and sponges, hydroids, sea-slugs, mussels and starfish live in water that is trapped and cut off for hours at a time. A rock pool may seem like a haven for animals and plants but conditions are not always as easy as it would seem—evaporation of water causes an increased concentration of salt, whilst a rainstorm has the opposite effect. Therefore, the species living within the rock pools are quite unique, being able to withstand changes in the temperature and saltiness of the water.

In addition to these three major subdivisions of the shore (which is often called the inter-tidal or littoral zone), is the sublittoral zone, which lies below the extreme low water level of the spring tides and represents the margin of what is truly marine. Plants and animals that inhabit this zone are never uncovered although they live in shallow water, which has a greater range of temperature than the sea.

Left: At the end of the day, the sun sets over a mangrove swamp in Australia.

Overleaf: Low lying coasts such as this one in Australia provide a great diversity of habitats for animals and plants, including sandy beaches, spits, sand banks, sand dunes and estuarine mud. The waves continually stir up the sand and bring a fresh supply of food to the animals living there.

The tides bring in all sorts of debris, including seeds such as the coconut (**Above**), which may germinate on the upper shore and grow into a new tree. Currents in the water are responsible for dispersing the seeds of plants growing on islands (**Left**).

Many birds gather at the water's edge, feeding on the wealth of animal life hidden in the sand. Flocks of black terns and black-naped terns (**Above Left**) feed on small fish carried in on the tide. A flock of black skimmers stand by small pool of water, watching for the slightest movements of fish or worms (**Left**). Dunlin and knot use their short bills to turn over pebbles (**Below Left**). Smaller birds, such as the tussock bird (**Right**) search through washed up piles of seaweed for sand hoppers (**Above**). Groups of ibis sift for worms and bivalves (**Below**).

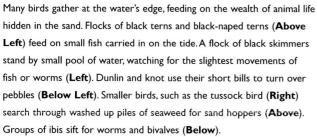

A sandy beach may appear to be devoid of animal life, but on closer inspection it is possible to find many animals buried in the sand. During low tide the animals stay hidden beneath the surface. The burrowing sea urchin (**Top Left**), or sea potato, uses its short spines for burrowing. Once below the sand, the animal covers the surrounding sand grains with mucus to create a protective chamber. These sea urchins are seldom seen alive, but when they die their bleached skeletons are often washed up on a beach. The sand mason worm (**Top Right**) cements together grains of sand to build a tube that sticks out above the surface of the sand. A ring of sticky tassels at the top trap food particles in the water, which the worm picks off with its tentacles. Herons search out whelks (**Above Left**) and razorshells (**Above**) buried in the sand.

Above: Very low on the shore, where there is nearly always a covering of water, fan worms build tubes that protrude from the sand or mud. Their crown of tentacles extracts food particles from the water.

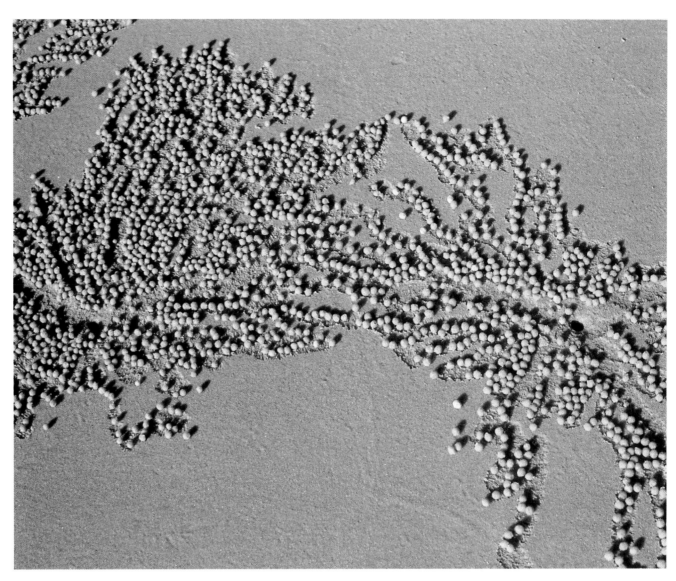

The ghost crab is common on tropical beaches. The mounds of sand on this beach in Australia at low tide are a clue to the number of crabs living on or in the sand (**Above Left**). These crabs are about five centimetres (two inches) across and have eyes on long stalks which gives them 360 degrees vision (**Right**). They build burrows in the sand (**Above**), often leaving characteristic patterns on the surface (**Top**). At night they appear to search the surface of the sand for organic refuse (**Left**).

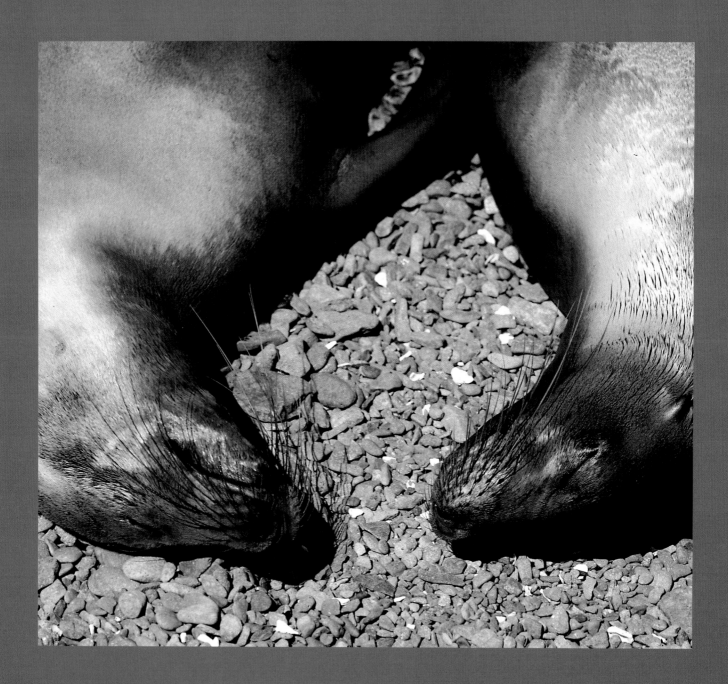

Seals and sea lions haul themselves onto sandy beaches to rest, mate and
to stay warm while they are moulting (**Above Left**). They are very ungain-
ly on land (**Above**).

While seals are not very graceful on the land they move with ease through the water (**Above**). The common seal (**Left**) is found on beaches of northern Europe.

The southern elephant seal (**Left**) prefers the beaches of the southern oceans. Elephant seals can weigh as much as 4,000 kilograms (9,000 pounds). During the breeding season, the males fight over the females, rearing up at each other and inflicting serious wounds (**Above**).

The female Weddell seals give birth to their pups on
the ice of Antarctica.

Shores that have a large tidal range expose vast expanses of mudflats at low tide (**This Page**). These provide a valuable habitat for flocks of shore birds such as the knot (**Top Right**). Thousands of animals live in the mud, including the sand plough snail which leaves a characteristic track across the surface of the mud (**Below Right**). The higher banks are colonised by plants such as the samphire (**Bottom Right**) which is adapted to living in salty soil.

Sand dunes provide cover for many ground-nesting birds such as the oystercatcher (**Top**). Few plants can become established in the continually moving sands above high tide level. One pioneering plant is the morning glory (**Left**). Its network of shoots cover the sand and prevent it from blowing away. The seeds of this plant are dispersed by ocean currents. The sea oat has a similar role, here it is growing on a white quartz sandy beach in Florida (**Above**).

At high tide, the incoming water surges down the channels that criss-cross a salt marsh (**Left**), bringing with it fish and prawns. At low tide, the channels are empty and covered with mud (**Top**). Salt marshes are important stopping-off points for migrant birds, such as grey lag geese (**Above**), which feed in the sea grass beds.

Previous Page: There are many species of mangroves, some no bigger than bushes and others that are tall trees of 25 metres (80 feet) in height. The spreading roots of this mature grey mangrove tree are striving to stabilise a tidal mangrove swamp in Australia. The grey mangrove is so named because its pale stems contrasts with its dark foliage.

In this closed tidal swamp forest, newly germinated seedlings can be seen in the foreground (**Above**). The arching prop roots trap silt and over the years the level of the mud rises (**Right**).

The stilt-rooted mangrove tree (**Far Right**) is surrounded by a carpet of spikes called pneu-matophores. They are vertical roots which stick out of the mud to absorb oxygen. As mud accumulates on the seaward edge of the swamp the mangroves advance and claim it with viviparous seeds, which germinate while they are still hanging on the branches. When they drop off the tree (**Above**) they quickly take root and grow into a new plant (**Left**). At low tide, periwinkles take refuge among the mangrove branches (**Right**).

Previous Page and Above: The salt water crocodile inhabits the
mangrove swamps and estuarine creeks of Southeast Asia and Australia. It
feeds on fish and birds but it can survive for a month without food.

Top: At low tide, mudskippers wriggle across the mud between the mangroves, using their two strengthened fore fins. They use the crabs' breathing technique of keeping their gill chambers full of water, and also take in oxygen through their skin. If they want to move at speed they curl their tails sideways, flick them, and shoot across the mud.

Above: The male fiddler crab has one greatly enlarged claw which he uses to signal to a female. If he is successful the female will come over and go with him into his burrow.

Left: The great egret waits in the shadows watching for the movement of fish or small invertebrates.

Above: Here, a blue hermit crab in an old whelk shell shelters in a bed of sea grass.

Left: The top predator in the mangroves of the Bay of Bengal is the tiger. A sub-group of the tiger has become adapted to living in the coastal region where they hunt small mammals and birds.

Far Left: The manatee is an endangered marine mammal that is found in swampy coastal regions, such as Florida, feeding on sea grass.

The ever-moving pebbles of a shingle bank make it impossible for plants to become established.

Right, From Top to Bottom: Further up the shore where the shingle is more stable, a few very hardy plants can be found (**Top**). These plants have to be tolerant of the salty spray and the near-drought conditions, for there is little fresh water. The thick, succulent leaves of the rock samphire (**Second from Top**), sea kale (**Third from Top**) and sea holly (**Bottom Left**) are more reminiscent of desert plants. The yellow horned poppy (**Bottom Right**) is an annual, growing for just one summer before setting seed and dying.

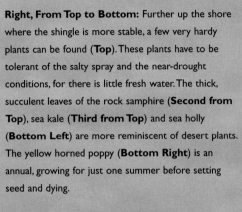

129

The ringed plover (**Above**) is ideally suited to life on the shingle banks.
Not only are its feathers well camouflaged against the pebbles, but its eggs,
laid amongst the pebbles are virtually impossible to spot (**Left**).

Kittiwakes (**Left**) nest in huge numbers on the narrow ledges of sea cliffs (**Above**). These birds spend much of their time at sea, fishing, but they have to return to land to nest and raise their chicks. Puffins (**Right**) feed on sand eels. They nest in burrows that they dig in the thin soil at the cliff top.

Overleaf: Like the kittiwakes, great flocks of gannets rear their young on shelves in cliff-faces.

A pair of puffins survey the nesting ground on a cliff-top high above the ocean.

Above: The well-known Seven Sisters are chalk cliffs along the south coast of England. As the cliffs are eroded, rocks fall onto the beach below and become a habitat for many inter-tidal animals and plants.

Above: The rocky shore is continually battered by the waves.

Magellanic penguins (**Left**) and blennies (**Above**) sit on the rocks above the high tide mark. Once they are warmed by the sun, the blennies return to the water to cool off. A xanthid crab takes refuge in a crevice during the heat of the day (**Top**).

Overleaf: Marine iguanas are similar to the blennies in that they warm themselves on the rocks during the day and enter the water to cool their blood when necessary.

141

The upper shore is only covered by water during high tide and when the water has receded clusters of small spike shells (**Top**) can be found in crevices. The blue chiton (**Above**) and the large tropical chiton (**Above Centre**) clamp themselves to rocks.

The sea slater (**Above Right**), the marine equivalent of a terrestrial wood louse, has a thick protective exoskeleton to reduce water loss.

Tiny insects such as velvet mites and thrips (**Right**) live among the rocks.
At the topmost level, beyond the highest limit of the most drought resistant wrack and the highest reach of the tides, where the only seawater to
arrive comes as spray, live tiny acorn barnacles (**Above**). They are so small
that they manage to sustain themselves on small particles contained in the
spray and they conserve the little amount of water they need within their
shells, as they are clamped to the rock.

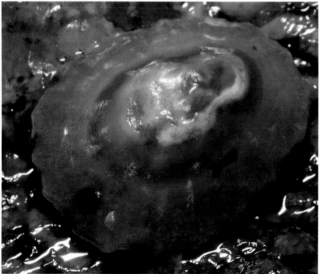

The middle shore has a much greater diversity of life. The tentacles of the beadlet anemone (**Above**) are fully expanded at low tide, but when it is exposed the tentacles are pulled in, leaving a red jelly-like blob on the rock. Limpets (**Left**) are able to move about and graze on seaweeds when the sea covers them, but when the tide starts to retreat they return to the rock and stick themselves down firmly to retain moisture during the driest part of the day. Mussels (**Top Right**) cannot withstand long exposure to the air so this determines the upper limits where they can be found. Their lower limit is fixed by the starfish (**Above Right**) that prey on them, but starfish cannot live long out of seawater so, several metres above low water mark, the mussels are able to dominate the shore. Mussels attach themselves to the rocks with bundles of sticky threads but their grip is not very secure so where the waves beat very heavily they cannot maintain their hold. The goose-necked barnacles (**Right**) then take their place, fastened very firmly to the rock by a thick wrinkled stalk.

The abundance of sedentary animals on the middle shore provides a ready source of food for the predators. The dog whelk (**Top**) feeds on mussels, barnacles and topshells (**Above**), by drilling holes through their shells with its many-toothed tongue, known as a radula. Crabs (**Left**) are active on the lower shore and in rockpools.

When the tide goes out water is left behind in the rock pools (**Above Left**), which form a refuge for the snakelocks anemone (**Far Left**) and sea urchin (**This Page, Top Left**) Occasionally, larger animals such as the squat lobster (**This Page, Top Right**) and the octopus (**Left**) may be trapped in a rock pool on the lower shore.

Fish are also found on the lower shore. The cornish sucker (**Above**) holds onto the rocks with a sucker beneath its head so that it is not swept out to sea by the waves. The feather duster worm (**Overleaf**) is a filter feeder, trapping particles of food on its tentacles.

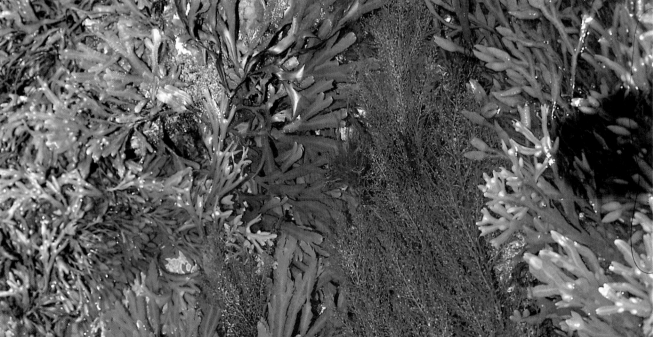

Seaweeds have become adapted to the inter-tidal zone by having thick, impermeable outer layers. Their resistance to drying out determines their upper limit on the shore. The channelled wrack (**Top**) lives on more exposed rocks as it can withstand drying out due to conservation of water in the channels along its fronds. The bladder wrack, with its characteristic air bladders for floatation, and the knotted sargassum are found as far up as the middle shore (**Above**). The toothed wrack (**Above Right**) is found nearest to the sea. The long brown oarweeds, Laminaria, (**Right**) grow profusely on the lower shore, projecting above the surface of the water at low water of spring tides. However, their holdfasts that secure them to the rocks are rarely exposed.

Left: In contrast to the Laminaria, green algae can be found in some of the brackish rocks pools near the high tide mark.

The sub-littoral zone lies below the lowest tide level. These shallow waters have a greater temperature range than the open sea (**Right**). Seals swim through these shallow waters hunting fish (**Below**).

DIFFERENT BEAKS
FOR DIFFERENT JOBS

To exploit the many types of food available birds have evolved into different species, each specialised for a different diet with the beak shape that is best suited to gather it. The diet may be herbivorous, carnivorous or omnivorous. This means that an estuary may be full of different types of bird but they are not competing directly for food. Some will be looking for fish while others are looking for buried shellfish. Some birds have short thick beaks, or bills as they are often known, for seed eating while others have hooked and powerful ones for tearing carrion. Some beaks are thin and pointed, typical of insect eaters, while others may be shaped like a spatula or spoon for scooping up food in water. Beaks are not only used for feeding, but also self defence, attack, nest building and even sometimes climbing; the length, strength and hardness of the beak varying considerably amongst the species.

The little blue heron (**Above Right**) has a long, dagger-like beak that is well suited to spearing prey. Herons feed on small fish, worms, and insects found in shallow waters. Like the heron, the curlew is a large wader (**Below Right**), but with a very long, slim, down-curved beak adapted for feeding on smaller animals, and even some berries in autumn. Its main diet though is insects and their larvae, spiders, small crustaceans, molluscs and worms.

Belonging to the same genus as the curlew is the smaller whim-brel (**Right**), with a shorter somewhat thicker and less decurved beak than the curlew. It inhabits flat, sandy and muddy coasts, small pools near the coast and also rocky shores, feeding on insects, snails, worms and crustaceans and, again, berries in season. The beak of the woodstork (**Below Right**) is similar in shape to the curlew, but it is much heavier and capable of dealing with larger animals such as fish and large crustaceans.

The dowitcher (**Below**) has a long straight probing beak to find worms and shellfish deep in the mud. The stilt (**Bottom**) is a small graceful wader with a fine, short, needle-sharp beak, living mostly on saline marshes and eating insects, small crustaceans, tadpoles and fish.

Living in muddy bays, lagoons and saltpans one can find a larger, elegant shorebird, the avocet (**Right**). It is unusual in having an upturned beak and, when feeding it sweeps its head from side to side to skim crustaceans, worms and aquatic insects from the surface of the mud.

On marshes and river deltas a very large, white waterbird is a familiar visitor. The pelican (**Below**) has a long beak with an elastic yellow throat pouch, very useful when catching and carrying fish. The spoonbill has a characteristic spoon-shaped beak, as its name suggests, for sifting and spooning up small fish, mussels, snails, tadpoles and aquatic insects.

A flamingo (**Far Right**) feeds by wading into the water on its long legs and lowering its neck so that its beak it held upside down in the water. Its tongue then shoots in and out like a piston, forcing water through a sieve in its beak that filters the water.

the O

PEN OCEAN

Now we really are going to get our feet wet! In fact we shall have to swim, snorkel and scuba dive, or even take out a submarine, to get a true picture of the world we are about to enter. We are moving away from the shore and into the vast expanse of water that extends out into the open ocean and plummets to depths of more than four kilometres (2.5 miles). How can we differentiate between the tidal zone and the coastal zone when there is so much overlap and, depending on the tides, so many of the organisms encountered on the shore are often also found in the open water? The answer is that nutrients flow into these regions from the rivers and also rise from the seabed, creating rich, well mixed water. It is not then surprising that the open ocean is the most productive part of the sea and very different from the variable environment with which plants and animals of the inter-tidal zone have to contend.

The seas are colonised by only one group of plants—the algae. Algae are simple plants, made up of unspecialised cells that show no division into root, stem or leaf, and nor do they flower or fruit like land plants. They range from single cells that make up the phytoplankton to the huge multicellular kelps. Seaweeds do not require a system for transporting food—the seawater provides all the essential elements they need to absorb—and they also lack the rigid cellulose skeleton seen in land plants because seawater also provides support. Marine algae need light, of course, just like any other plant and, as light does not penetrate very deeply into the water, seaweeds either float or attach themselves to the bottom where the sea is relatively shallow. In many ways they are very well adapted to a life in the open ocean.

Right: A vast expanse of open ocean that stretches for thousands of kilometres.

THE COLOURS OF SEAWEEDS

Seaweeds are divided into three main groups, roughly distinguished by colour—green, brown and red algae. All seaweeds have chlorophyll—a green pigment that is essential for photosynthesis—just like land plants. However, the green colour can be covered by another pigment, such as brown or red, and it is these light absorbing pigments that decide where the seaweed is able to grow.

As light penetrates the water it is absorbed according to how much sediment is suspended in the water. Some colours such as blue, penetrate much more deeply than others and as the water becomes deeper, plants tend to dominate on the rocks according to their light preferences. The large seaweeds, which are anchored to the seabed, are therefore restricted to areas where the bottom is sufficiently shallow for light to penetrate.

The green seaweeds (**Above**) absorb the long wavelengths of red light and reflect green, therefore they look green. They grow where they are exposed to the strongest light, in shallow waters, and store their food as starch like land plants. The brown seaweeds (**Above Right**) look brown because they absorb the medium wavelength green light. They are the largest and longest of the seaweeds and are able to live deeper because green light can penetrate further into the water than can red. They also show a preference for cooler water temperatures. The red seaweeds (**Right**) absorb the blue and ultraviolet wavelengths of light and red is the most common colour of seaweeds. They can live at great depths; even 600 metres (2,000 feet) is common in clear waters.

Kelps are very large seaweeds that can be many metres in length and are so prolific that their environs are described as kelp forests. They dominate the shallow zones, providing a major habitat for smaller plant species and hundreds of different types of animals, including fish, crabs, shrimps, starfish and sea urchins. The kelp fronds sweep back and forth with the movement of the tide and currents and as they do so they sweep rock surfaces. This sweeping action keeps the surfaces clear of sediment, which serves to encourage animals such as sea anemones and sponges to become established, and also other seaweeds to grow. Young kelp plants need clean rock surfaces on which to grow but once established on a rock the level of grazing by a variety of marine animals, particularly sea urchins, controls their numbers. Along coasts of the Pacific Ocean, feeding by sea otter populations in turn keeps down the numbers of kelp grazers, thus maintaining a natural balance.

In contrast to the huge kelps, phytoplankton float in the surface layers of the ocean. (The term "plankton" is a rather loose one meaning "that which wanders or drifts".) Despite their microscopic size, they are the most important form of plant life in the oceans, being the basis of all marine food chains. Apart from the over-riding requirement of sufficient light for photosynthesis, the two main factors influencing the production of plankton in the ocean are temperature and nutrient supply. As on land, the higher the temperature, the faster the biological processes, but in the ocean the main importance of temperature lies in its effect on water currents. If there is no mixing of the cold deep water and the warm shallow water there will be no supply of nutrient salts to the surface layers.

Left: Most seaweeds are anchored to the rocks by a disc-like holdfast. Some of the larger holdfasts are even the hiding places for many small creatures such as brittle stars and worms.

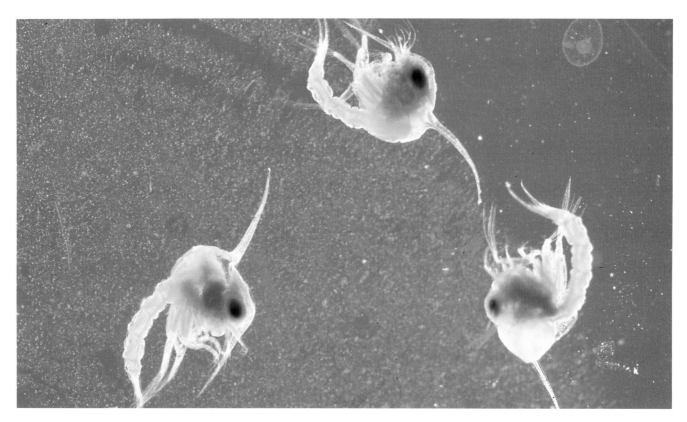

Above: Microscopic larvae of barnacles, crabs, prawns and lobsters make up much of the zooplankton.

Siphonophores (**Left**) are complex floating colonial animals that make up part of the zooplankton. They are related to jellyfish. The long tentacles of the jellyfish are covered with sting cells that are used to catch plankton, which show up as bright spots around the jellyfish.

With the right conditions phytoplankton multiply quickly. The greatest light intensity occurs over the tropics but nutrients, especially nitrates and phosphates are in short supply in these regions and this restricts growth. The most spectacular phytoplankton blooms are found in cooler waters where nutrients found in dead plant and animal waste are brought up from the bottom during storms. Nevertheless, upwellings of nutrient-rich water occur in both cool and warm waters and phytoplankton is found to a lesser or greater degree throughout the world's oceans. In spring, the onset of longer days and the plentiful supply of nutrients create a bloom of phytoplankton and a second, smaller bloom occurs in the autumn.

Phytoplankton are eaten by drifting animals called zooplankton, which contain representatives of most of the main animal groups, from the protozoans to larval fishes, and all sizes from microscopic forms to jelly fishes with discs a metre (three feet) or more in diameter and tentacles many metres long. Whether they feed directly on the phytoplankton or upon other animals, the zooplankton are ultimately dependant for their existence upon the primary food producers, the phytoplankton. The zooplankton therefore show a similar distribution to the phytoplankton, being most abundant in coastal water, in moderately high latitudes and in regions of high nutrients.

Unlike the phytoplankton, zooplankton are not restricted to the sunlit upper regions of the ocean. Planktonic animals are therefore found at all depths, though they are concentrated in the upper 1,000 metres (3,300 feet) or so. They themselves are food for small fish such as herring, which in turn are eaten by larger fish like dogfish, who are eaten by still larger fish or predators like dolphins. Strangely enough some of the larger ocean animals, such as whale sharks and blue whales, are involved in a much smaller food chain, feeding directly on zooplankton. When phytoplankton die they sink to the seabed along with gelatinous zooplankton remains, making sticky clumps called marine "snow".

Above: Kelp beds can be so thick that they can be seen as dark shadows in the water when viewed from the air.

Overleaf: The basking shark swims along with its mouth open, straining the water to extract plankton.

Fish are one of the many groups of organisms which populate the open ocean and they are usually the first type of marine animal that springs to mind when thinking about ocean life. From an ecological view point, the adaptations evolved by fish to life in the deep sea are fascinating and spectacular. They live at nearly all levels of the ocean from the air-water interface to the water-land boundary, and to the depths of the abyssal zone. Their differences in form and function relate to such survival pressures as concealment, feeding and reproduction. The majority of deep sea fish are found in the tropical and sub tropical regions where ocean clarity allows maximum penetration of sunlight.

In the well lit top euphotic layer, in the first 200 metres (650 feet) of the ocean, life is very different from the murky depths. It is populated by a wide range of fish, including the flying fish and the fast-swimming tuna, sailfish, marlin and swordfish. The sailfish is slightly longer than the tuna and dwells in warm temperate waters, often leaping clear of the water surface, making it a popular sport fish in some areas, whilst tuna is a popular fish for consumption all over the world.

The larger sharks also dwell in the upper zone of tropical and temperate waters. These streamlined fish grow up to four metres (13 feet), and are dark coloured on their ventral surface but much paler on their underside. Their saw-edged teeth develop from scales—which also cover their bodies—and as the teeth are lost they are replaced by new ones. As its dentition suggests, the shark is a voracious predator with squid and fish forming a substantial part of its diet.

Many species that live in the deeper ocean, though not on the ocean floor, rise to the surface during the hours of darkness and descend again as daylight approaches. The main stimulus for this so-called vertical migration is believed to be the need for smaller species to feed on the plankton near the surface. These small animals are followed by larger predators and these by larger still. Certain species may also rise to breed, and their young then contribute to the stock of zooplankton. The life in the upper sunlit ocean is therefore the key to the survival of the deep ocean animals. For while some creatures rise to feed, others are nourished by the continuous rain of particles and the descent of dead animals

Above Left: Many of the fish of the open ocean live in shoals. This gives them greater protection against predators.

Top: Flying fish have wing-like pectoral fins to allow them to glide over the water. A rapid burst of speed below the surface propels them into the air where they may glide for ten seconds or so, over some hundreds of metres.

Above: A tiger shark—the prefect predator with an excellent sense of smell.

Right: The tuna can grow to three metres (ten feet) in length and weigh up to 370 kilograms (800 pounds), yet it is the swiftest and most mobile of predators. It has a finely streamlined body and specially adapted fins to cut down on drag when swimming at speeds of up to 70 kilometers per hour (45 miles per hour).

Above: Grey seals swim in the coastal waters looking for shoals of fish.

Left: The bioluminescent light that flashes through the transparent skin of the comb jelly is produced by chemical reactions that take place in its eight digestive cavities.

Below Left: Some deep water fish achieve camouflage by becoming transparent.

Above Right: Turtles spend their adult lives at sea where they feed and mature. Every few years they swim great distances to return to their breeding grounds.

and plants from above, without which there would be insufficient food available. Thus there is an interaction and interdependence between the upper and lower layers of the ocean.

The twilight of the bathyal zone, between 200 metres (650 feet) and 2,000 metres (6,500 feet), not only has dim light but also correspondingly low temperatures. However, it has a diverse fish population, many of which are grazers and carnivores that make the nightly climb to feed in the upper well-lit zone. Their migration may span several hundred metres, and these are typically small fish of around ten centimetres (four inches) in length. Others reside permanently at the lower levels, feeding on migrant prey

entering or passing through their living depth. Exposure to predation is a major problem for creatures in the open ocean as such a habitat lacks shelter; this explains why so many fish species flock together in schools—a certain amount of security can be found in numbers.

The steady decline in the penetration of sunlight is also a main feature of this zone and many adaptations to the decrease in light are found: particularly in colour. The colours of mid-water animals are mainly related to concealment in the peculiar lighting conditions of the sea. Sunlight entering the ocean becomes more and more blue with increasing depth so, close to the surface, animals adopt a vivid blue coloration. On many fish, dark dorsal surfaces mean that they are camouflaged from above to sea birds, and a silvery ventral surface renders them inconspicuous to predators from below. Deeper still, fish become more heavily pigmented and are usually brown or black. Many species, both grazers and carnivores, also have upturned eyes to locate prey not so well protected.

Many animals of the bathyal zone and abyssal zone, over 2,000 metres (6,500 feet) deep, in the ocean, can emit light produced by the oxidation of special chemicals called luciferins and these can be used to lure prey. However, light organs may also be used for cam-

ouflage by lighting up the fish's underside, which would otherwise be seen from below as a dark silhouette. The larger, more elaborate light organs are generally found in the species living in the bathyal zone and they reduce in size the deeper the fish are found. Lanternfish, for example are about ten centimetres (four inches) in length and are so called because of the pattern of small light organs on their flanks and undersides. Hatchetfish are about the same size and live at mid-twilight levels; ideally suited to their surroundings these fish are flattened from side to side and thus present a minimal silhouette from below.

From the earliest times fish have been provided food for humans, and today this is no different. And while shrimps, prawns, crabs, lobsters and other types of seafood are also caught, fish are by far the most popular, with some 70 million tonnes caught around the world each year. They are caught in a number of different ways, by hand-thrown nets in local waters and at sea by modern fishing vessels with the latest technology. Some fish are caught on long lines with many hooks or ensnared in long walls of drift nets, while bottom dwelling fish are trawled, whole shoals being gathered up in huge nets set in mid-water. Using sonar to detect shoals means there are few places where fish can escape notice and there is great concern

at a world level that overfishing is causing numbers to decrease. Too many fish are being caught and the numbers cannot be replenished fast enough.

Ocean travelling species make the most of the vast expanses of water, criss-crossing the oceans to find the best places to feed and breed, often making use of currents to speed them on their way.

Perhaps the best known of the distance travellers are the whales, such as the humpback, which feeds in the food-rich waters of the far north or south and then travels to the warm waters of the tropics where the females give birth to live young. There are approximately 80 species of whales and dolphins all belonging to the same order and of these the majority are toothed whales. Among this group we find the small whales we call dolphins and porpoises, which measure only about one and a half to four metres (five to 13 feet) in length, as well as the 18 metre (60 foot) sperm

whale and the killer whales, beaked whales and pilot whales, measuring between four to nine metres (13 to 30 feet). The toothed whales feed mainly on fish and squid, which they pursue and then capture with their teeth. Even the mighty sperm whale eats squid, often catching them at great depths and it is not unknown for squid almost the size of the sperm whale itself to be eaten! Killer whales on the other hand may also eat the flesh of penguins, seals and dolphins.

Those whales that are not toothed are known as baleen whales, and the obvious difference between the two groups is their feeding apparatus. The baleen whales, instead of teeth, have a system of horny plates, called baleen, with which they filter or strain plankton from the sea. It seems unbelievable that the tiny plankton forms most, and in a few cases all, of the diet of these creatures as most baleen whales are very large indeed. Indeed, this group includes the

world's largest animal, the blue whale, which can measure up to 28 metres (92 feet) long, as well as the distinctive humpback whale, the right whale and the grey whale. Members of this group have been able to evolve to such a large size because water provides support for their enormous bodies. However, despite the size and weight of the larger members of the family, whales and dolphins are extremely mobile as they have evolved very streamlined bodies. Being mammals they must conserve body heat but they do not have a covering of fur or hair like land mammals. Instead they have a thick insulating layer of fat, called blubber, immediately below the skin. Whales and dolphins spend most of their life below the surface, some of it at considerable depths (the sperm whale can remain under water for more than an hour at a time), but, because they are mammals, eventually they must all come up to the surface to breathe.

Dolphins have long symbolised friendliness and intelligence and since the writers of classical Greece, dolphins have figured highly in human culture. It seems that they are even more popular today, when we are learning to respect these highly evolved animals, realising that it is unethical to treat them as performing circus animals and recognising them as true spirits of the open ocean.

Above Left: The classic view of a hump back whale diving under water after rising to the surface to breathe.

Above: Dolphins are naturally inquisitive animals and take pleasure in riding the bow waves of boats, giving us an excellent opportunity to watch them at close quarters.

A glimpse of the tail flukes of a whale breaking the
surface in the vast expanse of the Pacific Ocean.

The waters of the colder oceans always look grey and fore-boding. This is because the light is absorbed by all the particles in the water. The continually stirred coastal waters around northern Europe (**Above**) and the Southern Ocean (**Left**) are amongst the most productive water bodies in the world.

In contrast to the colder waters, the beautiful blue oceans of the tropics indicate that the water is free of life, and the light can penetrate to great depths. The aquamarine colour off the coast of Cuba (**Above**) and the Philippines (**Right**) show that the water is shallow and productive, but the midnight blue of the deeper water lacks life.

Kelp beds are found in coastal waters. Kelps (**Above**) are streamlined flexible plants that can cope with waves and currents. Some kelps extend their fronds as much as one metre (three feet) per day. The giant kelp growing off the coast of California can reach 100 metres (300 feet) in length, rising up from the seabed (**Right**).

Many different animals use the fronds as hiding places, including shrimps (**Above**) and wrasse (**Left**). Larger predators visit the kelp beds in search of food. The seal (**Right**) and European otter (**Above Right**) are fish-eating specialists and pursue fish underwater at great speeds. The sea otter (**Far Right**) prefers abalone, mussels and clams. It brings the shellfish to the surface where it floats on its back. Using a stone as a tool, it smashes open the shell.

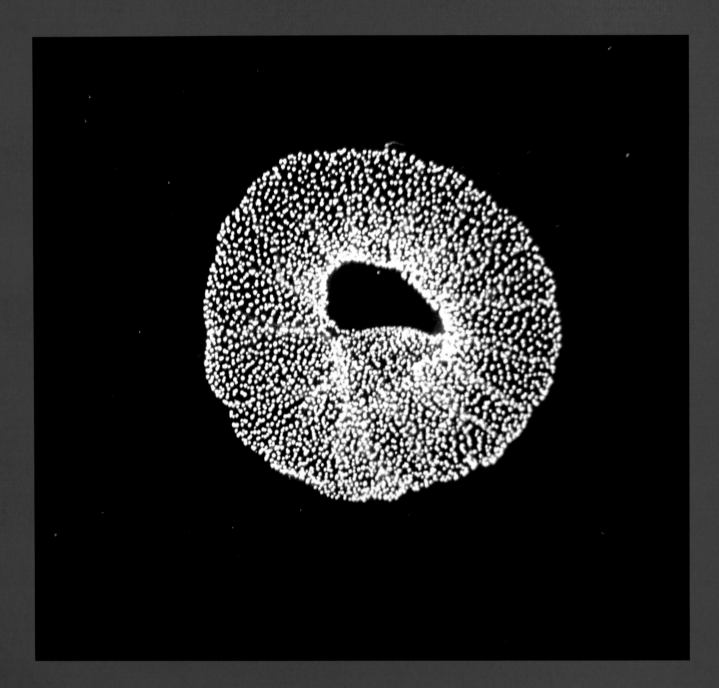

Plankton is made up of a very diverse group organisms, ranging both in size and form. The plant plankton, or phytoplankton, comprise two main groups; the unicellular algae called diatoms (**Left**) and the silicoflagellates. These two groups are responsible for 90 per cent of the plant productivity of the sea. Planktonic animals are drifters carried by the currents. They include the larvae of barnacles (**This Page**) and the jellyfish. The moon jellyfish (**Inset, Above**) hangs just beneath the surface, its weak movements preventing it from sinking.

The larger jellyfish (**Above**) can propel themselves forward by contracting their bell to force out a jet of water, but even so they are at the mercy of the currents. The lions mane jellyfish (**Above Right**) has bundles of tentacles that hang down in the water. These can paralyse small fish, which the jellyfish then consume. Baby squid also make up the zooplankton but once they are adult they will be able to move actively by jet propulsion. (**Far Right**). Ctenophores are pear-shaped animals with two tentacles (**Right**).

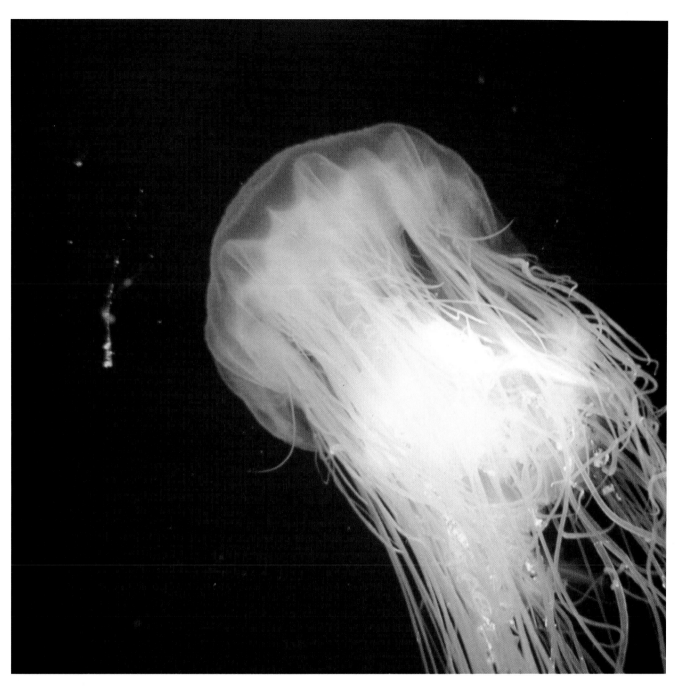

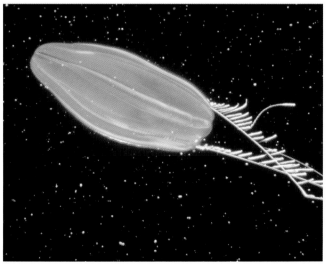

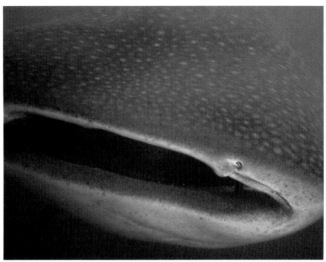

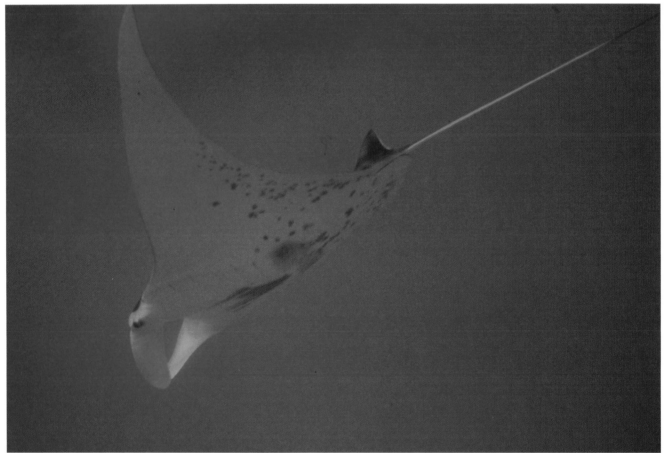

Filter feeders need to pass large volumes of water through their filtration systems in order to trap enough plankton to sustain them. There are ram feeders such as the basking shark, manta ray (**Above**) and anchovy (**Right**). These animals simply swim along with their mouth open. The huge whale shark (**Top, Left and Right**) is also a ram filter feeder. Some particles are small enough to slip through its sieve plate, meaning wasted food, so it coats the plate with a sticky mucus which can trap the smallest particles. The humpback whale (**Far Right**) swallows large mouthfuls of water and forces the water through its baleen or whalebone plates. These narrow plates hang like a curtain from its upper jaw, tapering down to divide into a fringe of bristles at the bottom end. Adjacent fringes overlap to form an effective filter. The southern right whale (**Above Right**), another filter feeder, is characterised by several large outgrowths or callosities caused by skin parasites on its head.

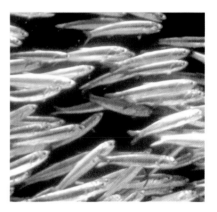

The open ocean is home to countless fish. Fish are ideally suited to a marine environment, having a streamlined body which is covered in scales (**Above**). The shoals of cod (**Left**) of the North Atlantic have great economic importance.

Most of the bony fish live in shoals, such as these snappers (**Above**) and sea bream (**Above Right**). The shoal moves and feeds as one, the largest being several kilometres across.

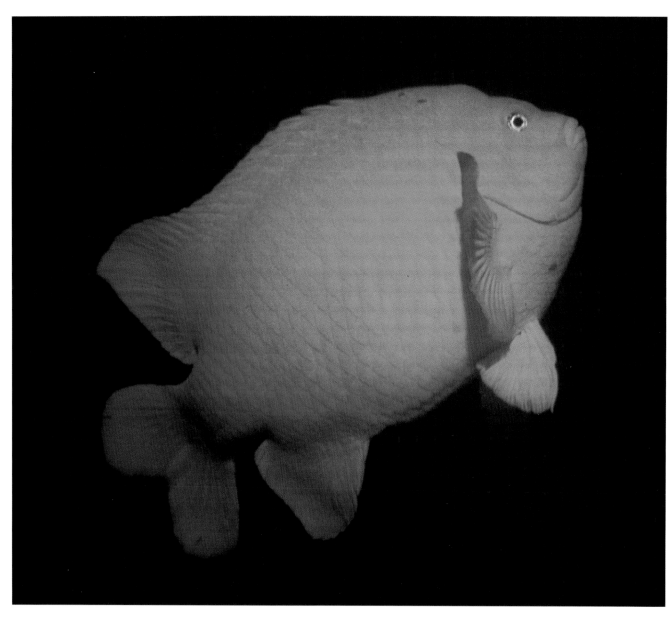

Above: The state fish of California is the brightly coloured garibaldi, which is found in the California channel.

Top: All that remains of a deep sea angler fish is its jaw bone. These fish have powerful jaws to catch prey and swallow it whole.

The shoals of fish attract the carnivorous fish such as the two metre-long (six feet-long) barracuda (**Above**) and tuna (**Overleaf**). The barracuda and tuna have an ultra-streamlined body and efficient blocks of muscle along their tails so they can speed through the water in the pursuit of small fish.

Spinner dolphins and yellow finned tuna hunt together in the Pacific.

The shark is a formidable hunter (**Above**). Its excellent senses, especially smell, enable it to locate its prey moving in the water (**Left**). Once located, it quickly swims towards the prey and grips it in its mouth (**Overleaf**). The backward facing teeth prevent the victim from escaping and once it has a good hold, the shark shakes the prey form side to side, a movement which helps its jaws and teeth shear the flesh into pieces (**Right**).

The most feared shark of all, the great white, grows to more than six metres (18 feet) and attacks any creature in the ocean, including humans and other sharks.

The dolphin (**Top**) makes use of echo location to detect prey such as fish
and squid. It generates clicks and whistles by trapping air and squeezing it
through a series of valves in its nasal passages. The sounds bounce off
objects ahead of the dolphin and the returning "echoes" are picked up by
a fatty channel in the lower jaw and transmitted to the ear. Many dolphins
travel in herds or pods and the members co-operate in the capture of
shoaling fish (**Right**). Killer whales (**Above**) also work in pods to herd
and trap salmon.

Right: The Weddell seal is the most accomplished of divers. It may be clumsy on land, but in the water it can dive gracefully to a depth of 580 metres (1,900 feet) in search of fish and squid.

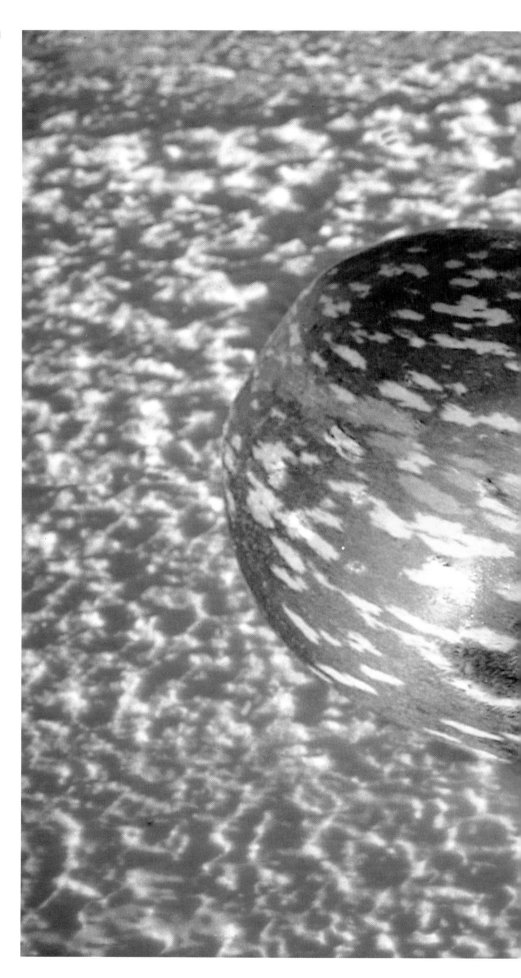

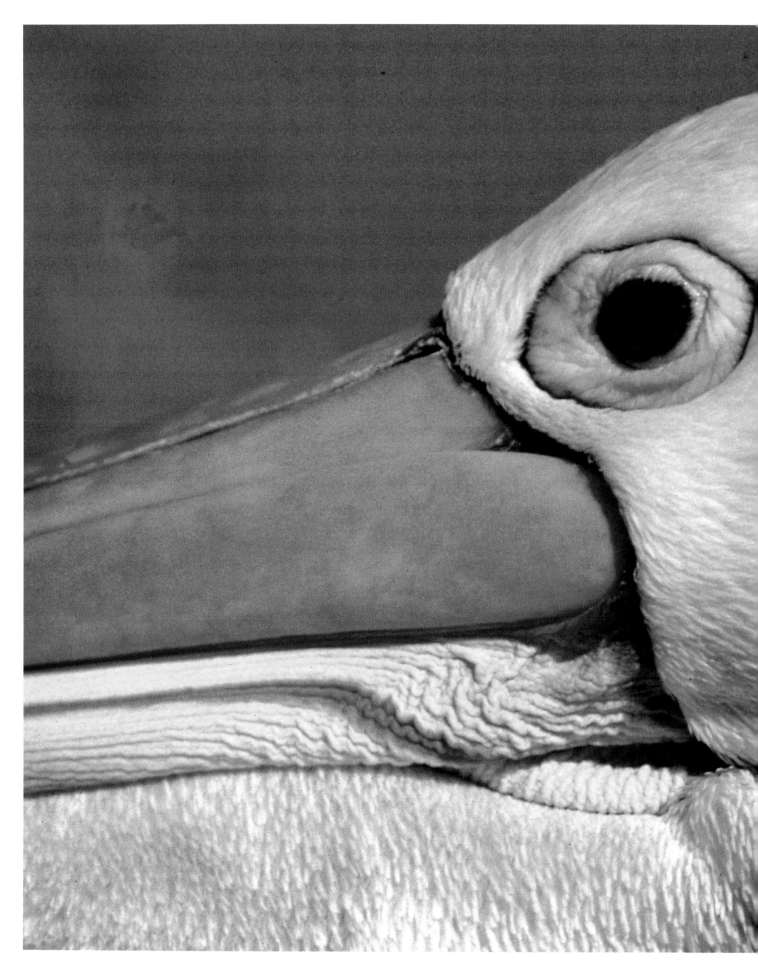

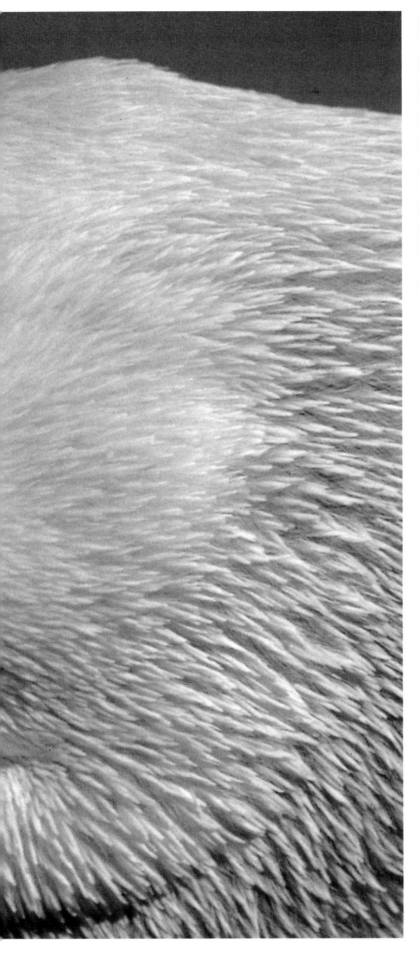

It is not just predatory fish and mammals that feed on the plankton-eating fish, birds too are attracted to this source of food (**Top**). Coastal birds fly each day from their nesting sites to the open ocean, including the pelican (**Left**) and the puffin (**Above**). The size of the puffin population is determined by the numbers of sand eel (**Below**), its primary food source.

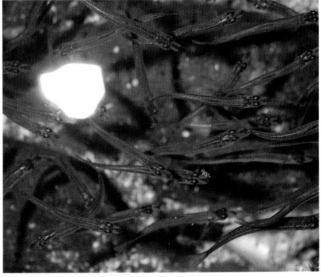

Like the puffin and the pelican, the gannet (**Left**) flies from its nest site to hunt over the ocean. Penguins, on the other hand, may be flightless, but they are superb swimmers. The wings have been modified into flippers which drive them through the water (**Top**). Their feet are positioned at the end of their body where they can be used for steering. This gives them their very upright stance when on land (**Above**). The flightless cormorant of the Galapagos (**Below**) also swims to catch fish.

The harvest of the oceans ranges from edible crab (**Right**) in coastal waters to squid (**Above**), the favourite food of the sperm whale. In Southeast Asia local fishermen catch a mix of fish and crustaceans which is sun-dried (**Left**). The cold waters of the North Atlantic supply Europe and North America with fish such as haddock (**Top**) and Dover sole (**Above Left**).

217

Many animals make incredible journeys across the oceans to breed or to search for food. The arctic tern (**Above, Far Left**) makes the longest journey of any animal. In autumn, it leaves its nesting sites in the northern most parts of North America and Europe and flies south 16,000 kilometres (10,000 miles) to its winter feeding grounds in the southern Atlantic and Pacific. It stays for the brief southern summer before journeying back to its breeding grounds again. Atlantic salmon spend up to four years at sea, feeding on other fish, and growing rapidly. Then the mature salmon make a hazardous journey swimming upstream (**Above Left**) as they return to their home rivers and streams where they hatched, recognising them by detecting particular quantities of substances dissolved in the water. Here they lay their eggs and leave their young to return to the sea and repeat the whole process all over again. Birds such as the cape petrel (**Right**) and the albatross spend their lives gliding over the oceans. The wandering albatross (**Left**) will spend two years on the wing, circling the southern oceans. Its large wings are

incredibly difficult to flap so the bird glides low over the ocean surface, making use of winds. It can fly for hours on end without having to flap its wings, swooping down to pick fish out of the water. However, it must return to its breeding grounds where it will mate and raise a single chick. The brown-browed albatross nests on the Falkland Islands, where it too raises just one chick (**Above**). The first flight for the young bird will be the most hazardous it ever makes as it has only one chance to take off from the cliffs and get airborne.

The baleen whales have to migrate between their feeding and breeding grounds. The oceans with the highest productivity are found close to the poles but these regions also show great seasonal differences. During the short polar summer there is a high density of plankton and this attracts the baleen whales, who may eat as much as four per cent of their body weight each day. They then return to the warmer waters of the tropics to give birth and rear their calves. The Californian gray whale (**Above Left**) spends the winter in the Arctic and migrates to Baja, California, for the summer. During the months of October and November humpback whales can be spotted along the Queensland coast of Australia. In the calving grounds, the whales are very playful and much rolling, fin slapping (**Left**) and breaching by both adults (**Above**) and calves is seen. The huge size of a whale's body can be appreciated when observed underwater (**Right**).

The enormous green turtle (**Above**), weighing in at 400 kilograms (900 pounds), lives in warm waters in the Atlantic, Pacific and Indian oceans. It may spend as long as 50 years at sea feeding and swimming, but like all turtles it has to go ashore to lay eggs. Some astonishing green turtles are known to feed on the algae and eel grass in one part of the world and then travel several thousand kilometres or more to reach their breeding beaches, where they themselves hatched (**Overleaf**). For those turtles living in the seas around Indonesia it is a 2,000 kilometre (1,250 mile) swim to the beaches of Queensland, Australia. First they mate in shallow water (**Above Right**), then the female comes ashore at high tide. She crawls up the beach to dig a large hole in the ground (**Far Left**) into which she lays as many as 100 leathery white eggs, the size of billiard balls (**Left**). She then covers the eggs

with sand and heads back to the water, returning several times in one breeding season to lay further batches of eggs (**Right**). Several months later, at full moon, the hatchlings break out of their eggs. The hatch is synchronous so that all the tiny turtles emerge at the same time and then have to make a desperate dash to the sea (**Far Right**). Unfortunately most fall prey to the hoards of gulls, rays and crabs that gather at this time, anticipating a feast. A few, however, make it to the open sea where they disappear under the waves. They will spend many years feeding and growing before making the return journey.

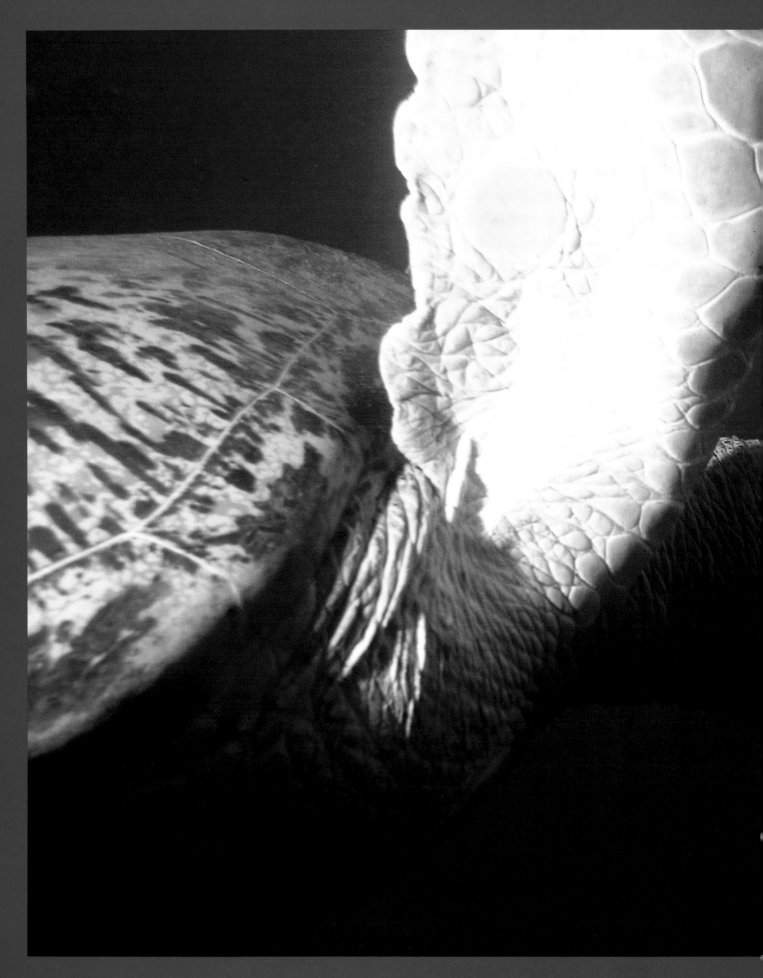

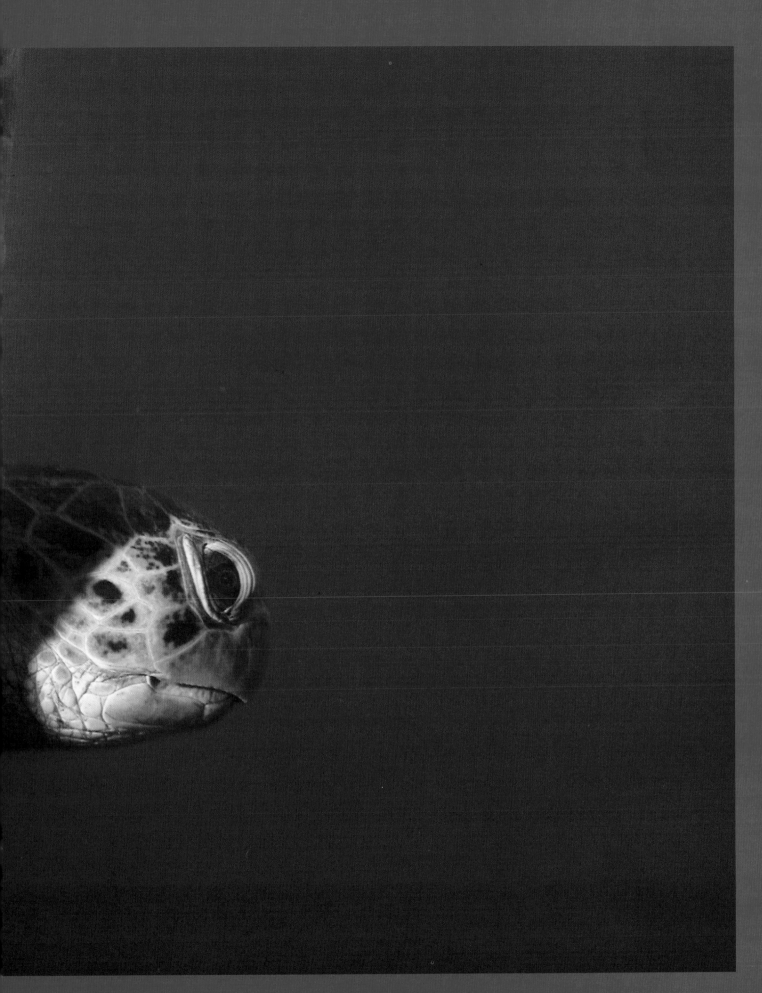

DOLPHINS

There is something magical about dolphins, and they have even been considered as the animal to symbolise the new millennium. We see them as intelligent beings like ourselves but with extraordinary powers. Like us they breathe air and yet they can survive in the wonderful underwater world that we can only glimpse. They seem graceful, peaceful, free, energetic, fun-loving and even appear to possess healing powers. We often forget though the other face of dolphins; as predators in their search for food.

Twenty-seven species of oceanic dolphin have currently been identified and they live in seas, oceans and many major rivers around the world, from the warm waters of the tropics to the cold waters of the poles. The most familiar member of the dolphin family is certainly the bottlenose dolphin (**Above**), with its well-known smile, but there are dolphins of many different colours, shapes and sizes ranging from 1.2 to 3.8 metres (four to 12.5 feet) in length. Some live close to the shore, while others spend their whole lives far out in the ocean; some are very active at the surface and spend a lot of time leaping out of the waves to a great height (**Right**), while others spend more time quietly swimming. There are even different feeding habits, though the majority feed almost exclusively on fish and squid.

Dolphins evolved from terrestrial mammals millions of years before humans emerged and they have become superbly adapted to underwater life. Their body shape has become streamlined (**Above**) and they have lost most of their body hair, their front limbs have turned into flippers and their hind limbs have disappeared, with their muscular tails providing a powerful means of propulsion. Perhaps the features that amaze us most are the blow hole (**Right**) through which the lungs are replenished with air, the amazing method of echo-location whereby they build up a "picture" of their surroundings with the help of sound, and the clicks with which they communicate. The marine environment is so different from our own and dolphins have perfected some incredible ways of navigating, communicating, finding food and avoiding predators.

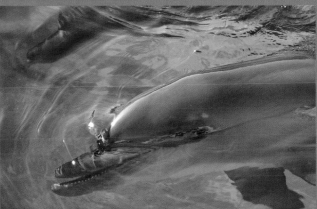

HAMMERHEADS

Hammerheads (**Above**) are one of the most fascinating and highly evolved types of shark. The front of the head between the eyes of all sharks is spanned by "Ampullae of Lorenzini"—sense organs that detect changes in pressure and electrical field. In the case of hammerhead sharks, the width of the head makes these senses more acute than those of other sharks. The head also acts as a wing, improving manoeuvrability and gives the wide separation of the eyes which means superior vision. The largest type of hammerhead is the great hammerhead, with a wide flat front to the head, and this kind grows up to 5.5 meters (18 feet) long. Great hammerheads are usually solitary and have been known to attack humans.

Unique amongst sharks is the shoaling behaviour of the scalloped hammerhead. Most sharks are solitary or co-operative in small groups, but scalloped hammerheads come together in enormous shoals of hundreds of individuals to mate (**Above, Far Right**). Female scalloped hammerhead sharks grow up to 3.5 meters (11.5 feet) in length and the males are smaller than the females, only reaching about two thirds the size of a fully grown female. A shoal of several hundred can occupy an enormous volume of water. The females manoeuvre for position in the middle of the shoal, with the biggest occupying pride of place in the centre (**Right**). By

being in the dominant position, the females make themselves more attractive to the males. At dusk, males and females will pair off for mating. When mating, the male wraps his body about the female and holds on with his teeth. Many of the large females in a shoal will bear obvious mating scars on their sides as a consequence. Individuals and small groups will break off from the shoal and swim in to the reef to be cleaned of parasites. When ready for cleaning they stop swimming, turn on one side and hold still while wrasse and angelfish peck away at their skin (**Right**). At night the shoals disperse while the individual sharks hunt for food. As the day breaks, the sharks return to re-form the shoal and the whole dance is repeated.

the Coral Reef

Imagine that you are sitting on a pale, almost white, sandy tropical beach. The sun is burning down from a cloudless, blue sky and you look out over a turquoise green sea. In other places the ocean may be shades of blue or grey but here it looks green. You will notice that there are large areas where it looks dark and almost forbidding, but on closer observation you see that in fact this is a very popular area of the sea and there are people snorkelling excitedly. This is because here there are coral beds.

Diving on a coral reef is an incredible experience. The sensation of moving freely in three dimensions in the clear sunlit water, in which corals grow, is awe-inspiring. You are in your own peaceful world, silent except for your own breathing and the crunching of the brilliantly coloured parrot fish as they graze on the coral with their beak-like mouths. Neon-striped grunt and pink-tinged snapper swim past in schools, their golden bodies sparkling under the sun's spotlight as they turn together on cue. A troupe of tiny silversides may part in unison to let you pass. Here even the largest of marine creatures can appear most graceful and you may be lucky enough to see turtles gliding effortlessly through the water, majestic rays soaring like birds in flight and large harmless nurse sharks prowling across the ocean floor.

Continued on page 236

Left: A Caribbean island with sandy beaches and coral beds lying offshore.

Right and Overleaf: A coral reef looks like an underwater garden with structures, in all shapes and sizes, that waft gently with the current.

CORAL GEOGRAPHY

Corals grow best in clear water with a temperature between 20 and 30 degrees Celsius (70 and 85 degrees Fahrenheit), with the optimum temperature being about 24 degrees Celsius (75 degrees Fahrenheit). These environmental factors put a limit on where coral reefs can be found, but in the Indian Ocean, Southeast Asia, Central Pacific, Southwest Pacific and Caribbean conditions are perfect and coral reefs can be found in abundance. Cold currents on the west coasts of the African and American continents limit coral in these waters to a few locations close to the equator. Off Ecuador, the Galapagos archipelago is the site of many conflicting ocean currents. Situated right on the equator, on some islands waters are cold enough for colonies of penguins to survive, while on other islands just a few tens of kilometres away the water is warm enough for corals. Silt in the water cuts out sunlight and hence constrains the rate at which corals can grow. Therefore on the east coast of South America, murky water from the Amazon estuary limits coral growth in an area that is otherwise within the preferred temperature limits. There are even a few exceptions where corals survive in colder water, such as the northern Red Sea and eastern coast of Florida, where winter water temperatures can drop below 18 degrees Celsius (64 degrees Fahrenheit). Also, in some locations corals survive in water as warm as 33 degrees Celsius (90 degrees Fahrenheit), the Persian Gulf and northern part of the Australian Great Barrier Reef being examples.

Above and Right: Clear, warm water is ideal for corals. They are only found in relatively shallow water where there is plenty of light.

Just as the rainforests are composed of an enormous variety of plants and animals that depend on each other for food and shelter, so are the coral reefs. Here too there are many different species living in the varied habitats created in the different layers of the coral "rainforest of the sea". There is, however, one very great difference between a forest on land and a coral reef; the first is formed primarily by plants and the second by animals. However, you could be forgiven at first sight though for thinking that coral was also a plant.

The coral reefs are an ancient habitat—many having been in existence for more than 500 million years—and they are exquisitely beautiful places. There are coral domes, like the huge brain coral, branches, fans, antlers delicately tipped with blue, and blood-red organ pipes. There is a brilliance of colours ranging from purples and blue, to fiery red, yellow and orange. Living in and around the corals are millions of fish and other animals, all dependant in one way or another on the coral reef and each other for their existence.

Above: A diver swims across the surface of the coral reef in the company of fish.

Above Right: A turtle swims over the corals in search of food.

Right: A coral is made up of tiny polyps, which are all connected to form a colony. Each polyp looks like a miniature sea anemone.

Coral reefs are the largest structures built by any animal on the planet, humans included. Compared to these limestone structures that are millions of years old, thousands of miles long and weigh billions of tons, human cities are a small and recent challenge. The amazing thing is that a coral reef is the cooperative result of many lifetimes of work by minute organisms—the individual coral polyps. Corals are colonial animals, with each individual polyp an animal in its own right. The coral polyps secrete their skeletons from their bases and each is connected to its neighbours by strands that extend laterally. But coral polyps cannot survive alone, even in their colony, for they are very demanding in their environmental requirements. Water that is muddy or

fresh will kill them and they cannot grow at depths beyond the reach of sunlight for they are dependent on single-celled algae that grow within their bodies.

The relationship between the coral polyps and these zooxanthellae algae, contained within the cells of the coral, is described as mutualism, which means that two or more organisms are interdependent, working together to live and grow. In the case of corals, the algae photosynthesise, just like other plants, to make food for themselves, and in the process absorb carbon dioxide from the water. This assists the corals in the building of their skeletons and releases oxygen, which helps the corals to respire. The coral polyps also feed on the zooxanthellae algae and on nutrients filtered from the surrounding seawater. In turn the algae feed on the waste products from the coral polyps.

As we have described earlier, the colours also vary considerably, though they are primarily the colour of the mutualistic zooxanthellae algae, cream and brown being two of the most common. One of the most important factors affecting colour is depth, as water absorbs more light towards the red end of the spectrum. In deeper water, therefore there is proportionately less red light and more blue light.

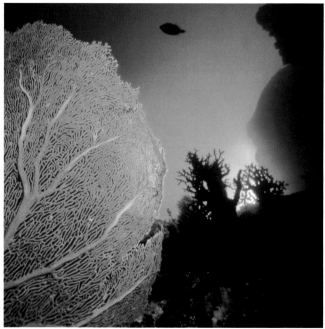

Top: The coral polyps vary in shape and size, the maximum size being approximately that of a small flower such as a daisy and minimum size about that of the head of a pin.

The huge fan-shaped gorgonian corals (**Above**), a soft coral, are found near the surface where there are currents of water. The deeper water coral species tend to be round boulders (**Right**) and overlapping sheets. These structures still give a good surface area for polyps to grow in sunlight, but are more robust and better suited to slower rates of growth.

As coral polyps grow, reproduce and die, their limestone skeletons grow on from those of previous generations. Over tens of years these build into structures in a multitude of shapes that we commonly think of as corals, but it is only the polyps on the surface of the coral that are still alive. Inside the coral is just dead limestone. Coral growth and reef life is most prolific just below the surface where the sunlight is strongest and the top few meters of a coral wall frequently grow outwards to overhang the older corals below. Shallow water corals close to the surf zone typically build thorny branching structures to obtain the maximum surface area for polyps to grow. With abundant sunlight for photosynthesis and a regular washing with fresh water, these corals have relatively rapid growth, some species growing 30 centimetres (12 inches) or more in a year. Natural breakage by storms is all part of the life cycle of these corals, with broken branches settling to the reef and growing into new corals. In fact, breakage happens even without storm

Above: A blenny hides in a crevice in the coral reef.

Right: The parrot fish works head down at the coral, bumping and scraping the surface with its beak. It feeds on the dead coral and algae that cover its surface; thus parrot fish keep the whole of the reef clean of algae. Having fulfilled this function they egest the remains of the coral in their faeces. There is so much that a great deal of the white sand that covers coral beaches comes from the gut of the parrot fish originally!

damage as left to grow the branching corals will eventually expand beyond the ability of their limestone skeletons to support them, collapsing under gravity.

The reef-building corals are described as stony corals because of their limestone skeleton but there is another type of coral, the soft coral. Soft corals produce a yielding, flexible skeleton that holds the colony of polyps together. All individual coral polyps in a colony, whether soft or hard, are connected by a thin skin-like layer that stretches over the skeleton.

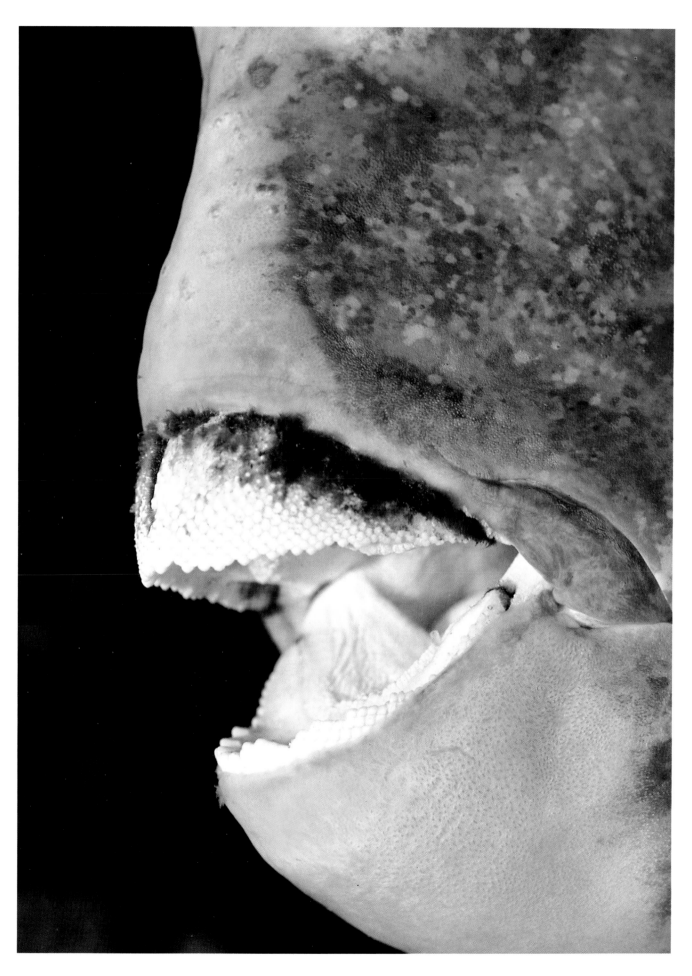

Over hundreds of thousands of years, limestone coral skeletons build to give the overall reef structure. Although appearing solid from a distance, the surface of a coral reef is actually a complex honeycomb of small tunnels; the gaps left between growing corals remaining as the reef grows upwards and outwards. These provide natural hiding places for many of the fish and crustaceans that inhabit a coral reef. Further behind the face of the reef, many of the cracks are filled with limestone sand, eventually solidifying to become part of the overall structure.

The zooxanthellae algae living in mutualism with the coral polyps are not the only algae on the reef. Where sufficient light is available other algae grow as small plants on the coral and many of these secrete limestone as a product of photosynthesis, giving them a semi-rigid frame. Debris from these calcifying algae provides sand to help fill in the overall reef structure. Other non-calcifying algae

grow as a fine covering over dead coral, and the parrot fish have evolved specifically to graze on these. Rather than lips and teeth they have hard "beaks" for crunching the surface of the limestone and any small algae growing on it. The algae are digested and the limestone is excreted as yet more fine sand, settling to the reef and accumulating to fill holes. New live corals can have a problem becoming established where previous dead corals are overgrown with algae. However, by grazing the algae, the useful parrot fish keeps the surface clean for more coral to grow.

Coral polyps secrete a layer of mucus for protection from bacteria, algal growth and grazing fish and normally this is a perfectly adequate sageguard. However, they are very vulnerable to damage by humans. The slightest graze by a fishing net or passing diver can damage the protective mucus and whilst the damage may not be immediately apparent, eventually the break in protection can lead to polyps becoming diseased or eaten, or algae growing on them.

The coral reefs grow over geological timescales and it is these enormous periods of time that lead to the various overall forms of coral reef. Over geologically significant periods, land moves, temperatures change and sea levels vary. Consider a volcanic island rising out of the deep ocean, a simple cone shape descending into deep water. Spores from corals and other marine organisms arrive as plankton floating in the current and settle on the rocks. As corals start to grow on the volcanic rock a fringing reef ecosystem becomes established on the slopes below the water. A fringing reef, typically, has several zones. The back reef is the zone nearest the land where the water is shallow and filled with coral debris. The reef crest is the area where the coral almost breaks the surface and waves break on the reef. This is usually too rough for coral so there are encrusting coralline algae. The reef front is where the corals can be found in abundance. Those growing at the top of the reef are more tolerant of the rough conditions while the more delicate forms are found in deeper water. Shallow water corals grow faster than the deeper corals, upwards and outwards to build a reef that is steeper than the underlying slope, eventually creating a vertical

Left: A vertical wall forms at the edge of a fringing coral reef.

Right: Some divers find it tempting to touch the corals, but the slightest touch will remove the protective mucus that covers the polyps.

Once a year, at the time of a full moon, the corals release their spores into the water. The synchronised spawning means that there is a greater chance of some of the spores surviving.

The fringing reef around Bora Bora in Tahiti (**Above and Below Right**) has grown out from the island. Behind the coral reef is a shallow channel. The reef front is clearly visible, being the point where the colour of the water changes from aquamarine to deep blue. Heron Island in the Great Barrier Reef (**Left**) is a coral cay, a low sandy island forming where sand builds up by wave action.

Above Right: On the reef, within large caves and beneath overhangs, dense shoals of glassfish behave almost as if they are a single blob of fluid suspended in the water.

wall that can even overhang in places. It is an extraordinary experience for divers to be floating over a sunlit coral reef when suddenly they reach the edge of the reef and all below them is inky blackness for thousands of metres. If sea level drops or land rises, corals above the surface die to form a limestone cliff, while below the surface the reef continues to grow outwards and upwards to fill the increased column of water.

Although just as close to the surface, coral nearer to the land is sheltered from the food-carrying ocean currents and cannot grow as rapidly as coral at the top of the seaward wall. Eventually then, a fringing reef slowly grows away from the land to become a barrier reef. In extreme cases, the original land drops entirely below the surface of the sea to leave an atoll reef still growing upward and outward. An atoll is a round or horse-shoe shaped structure with a central sheltered lagoon. Most atolls begin as a fringing reef around a small island; the island slowly sinking, allowing the coral's upward growth to keep up with the sea level. As geological time progresses, the sea level may stabilise for a period or even drop. Sand accumulates on the atoll reef, piled up by wave action and secured by vegetation and new islands become established above the coral limestone foundation, while the reef continues to grow upwards and outwards. Along main coastlines, a similar process forms fringing reefs that slowly build their way out to sea. Changes in sea level influence coral growth to produce barrier reefs along entire coastlines, which are separated from the land by a lagoon or wide channel.

The corals form both the basis for the food chain and provide accommodation for a plethora of reef dwelling creatures. Teeming clouds of small fish surge in and out of the reef in chaotic formation, pecking at minute specks of plankton in the water. Within the shoals individual fish appear to move at random, it is only when a threat is detected that the entire shoal acts as if with a single mind, to dart into the reef and hide. The danger gone, one or two fish tentatively poke their heads out, then again move out *en masse* to resume their random pecking at the plankton. Even without the reef to hide in the close packed shoal, and random behaviour within it, serve the purpose of confusing any attacking predator. Some shoals even comprise a number of different species of a particular fish, co-operating to gain safety in numbers.

Reef dwelling predators often hunt by stealth. Scorpionfish and stonefish, for example, are remarkably well camouflaged, lying amongst the coral rubble and sand. They surprise their prey in a split second, during which time they can suck in their victim with a large gulp of water. Although depending on camouflage for predation, it is not their only defence. Scorpionfish, stonefish and their relative the lionfish, all have spines in their dorsal and pectoral fins. When discovered and threatened they will flare these fins out and turn away to place the spines between themselves and the threat. The uncovered fins are brightly coloured to warn predators to keep away or face the consequences of pressing the attack.

Amongst the top predators are barracudas, long thin fish built for short bursts of amazing speed and ferocity. On the reef large barracudas will be solitary or found in small groups, whilst younger barracudas often form enormous spiralling shoals.

Left: In the Red Sea, bright orange anthias, looking a bit like swarms of common goldfish feed on plankton. In other reefs, various species of anthias and chromis have colours varying from yellow to purple, with a plethora of striped variations in-between.

Overleaf: The scorpion fish has spines, but these are not poisonous. It hunts by stealth, lying on the coral and waiting for prey to pass close by.

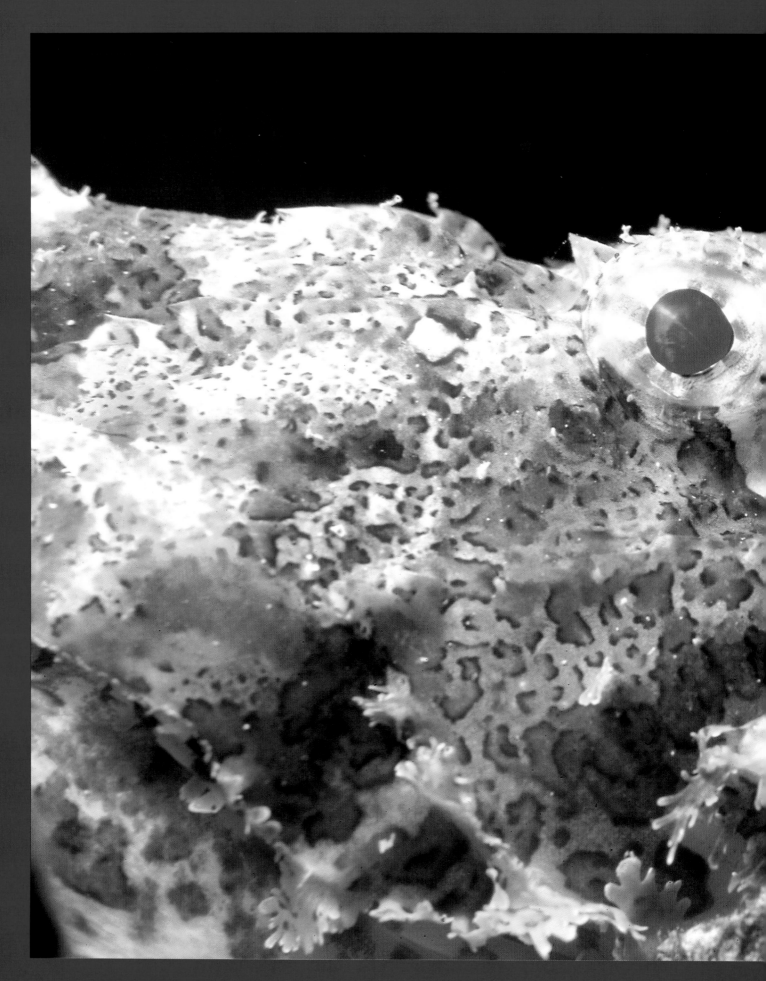

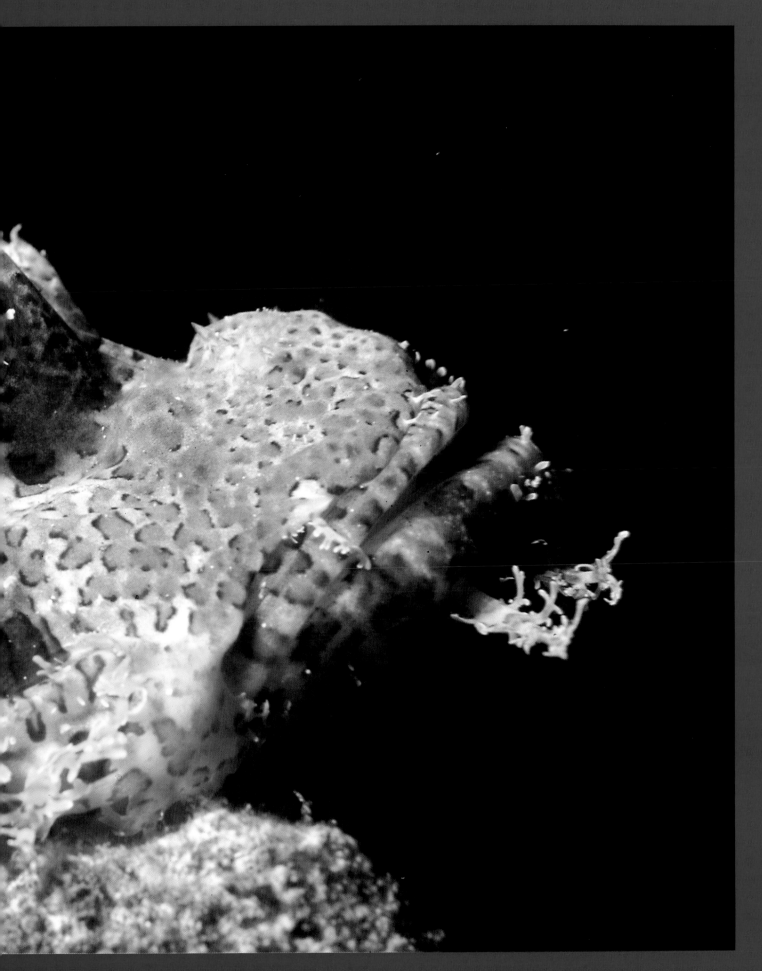

Of course, the most fearsome predators of the reef are the sharks. Most sharks are superbly efficient hunters and their only natural threat would be a larger shark. The feeding behaviour of hunting sharks can be likened to that of many large land predators, for they tend to hunt creatures smaller than themselves, feeding preferentially on any fish or other marine animal that is injured or otherwise vulnerable. They will not attack animals approaching their own size, unless the animal is obviously in distress, for being a successful predator they will not risk injury from prey that may well be able to defend itself. Only a few very large sharks are openly predatory towards humans. Tiger sharks and great white sharks are the main culprits for shark attacks on bathers and surfers. However, divers have been known to swim quite safely with these fish in circumstances where their feeding behaviour is not triggered. It is when vertical at the surface, when divers resemble an injured animal, that shark attacks on divers are most likely to occur.

Closely related to sharks are rays. They are both elasmobranchs, fish characterised by cartilaginous boneless skeletons, but unlike sharks rays are entirely harmless. Like many species of shark, the largest rays are filter feeders on the plankton of the coral reef.

Left: Pufferfish have a unique ability to inflate themselves several times their original size. This capability can be used to jam their bodies in a hole where a predator cannot get at them, or to expand to the size where they are simply too big for a predator to bite.

Above Right: Common on most coral reefs are white-tip reef sharks, which are fairly thin sharks growing up to two metres (6.5 feet) long, with a white tip on their fins. White-tip reef sharks often rest in caves beneath overhangs or gullies on the reef, only bothering to move if disturbed. They hunt small fish and a variety of other reef creatures.

Right: Reef stingrays can grow up to two metres (6.5 feet) across. When not hunting for food they bury themselves in sand with just their eyes showing. Their name comes from a venomous barb that is normally folded flat along the top of their tail, but should anything try to bite the body of the ray, the tail whips over and spears them with the barb.

Fish are not the only animals found on the coral reef, for it is an incredibly diverse ecosystem with inhabitants adapted to fill every available ecological niche. Nudibranchs are shell-less marine molluscs and their name quite literally translates from latin to "naked gill" for they have delicate external gill structures on their backs. Individual species are highly specialized for grazing single sources of food, usually sessile organisms (that is, an organism which is permanently anchored to its spot and does not move). Octopus and cuttlefish are much more highly evolved molluscs, able to change texture and colour at will to create near perfect camouflage against any background, a skill they use both for hunting and survival.

Sponges grow in many different forms including barrels, tubes, fingers and amorphous blobs, all encrusting the surface of the reef and even burrowing in to it. The various shapes have all evolved to facilitate efficient filtering. Again a colonial organism, sponges actively pump and filter water passing through their structure.

During daylight most corals polyps stay retracted within their limestone skeletons for protection from grazing fish. It is only at night while their predators are sleeping that they fully extend, covering the surface of the coral with living tissue, while the polyps filter plankton from the surrounding waters. The hard skeleton is often fully obscured to give the coral an altogether different appearance to that in daylight. As darkness falls, most reef fish nestle deep inside cracks in the coral, safely tucked away from hungry night-time predators. Indeed, most of the reef fish seen in daylight find a sheltered spot in the reef to hide at night.

An exception is the moray eel, which stays secure in a hole in the reef during daylight but at night ventures out to hunt fish and crustaceans. During daylight hours moray eels strangely often share their homes with spiny lobsters, one of the very crustaceans they hunt at night! Many other crustaceans also only venture out at night, often with their own portable protection. Hermit crabs carry discarded seashells that they can hide inside when threatened while anemone crabs decorate their backs with living anemones. The anemone benefits from movement through the water from which it filter-feeds, while the crab is protected by the stinging cells of the anemone.

Some of the most distinctive reef creatures are the anemone fishes, the most striking being the orange, black and white splotched clown anemone fish, though there are many colourful variations. As the name implies, anemone fish live in anemones and are protected from the anemone's stinging cells by mucus on their skin that carries the host's chemical signature. The first response of this fish to any other coming near its anemone is to dart out aggressively, in an attempt to drive the other fish off (**Above**). Only when all that aggression and posturing fails will it actually retreat to the safety of its anemone, hiding amongst the tentacles where it is safe from the stinging cells; but the attacker is not (**Above Right**).

Right: Many nudibranchs have beautiful colours and this advertises that they are poisonous.

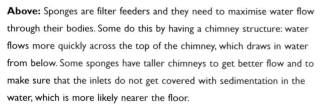

Above: Sponges are filter feeders and they need to maximise water flow through their bodies. Some do this by having a chimney structure: water flows more quickly across the top of the chimney, which draws in water from below. Some sponges have taller chimneys to get better flow and to make sure that the inlets do not get covered with sedimentation in the water, which is more likely nearer the floor.

Above Right: Night has fallen on the world of the coral reef and many divers find this the best time to explore the wonders of the ocean.

Right: The parrot fish finds a safe hiding place to rest during the night.

Far Right: The moray eel is a predator, grabbing unaware fish as they swim past its hole.

The temperature of the water around the Solomon Islands (**Top**) and the
Diego Garcia Archipelago (**Above**) is ideal for corals.

Left: At low tide the coral is just below the surface of the water.

Above Left, Above and Overleaf: Staghorn corals have a branching
shape that gives more access to space and allows all the polyps good
exposure to the currents carrying their food. These corals grow in abundance at the shoulder of a reef where the water is well aerated, but they
can be broken off their stalks by stormy weather.

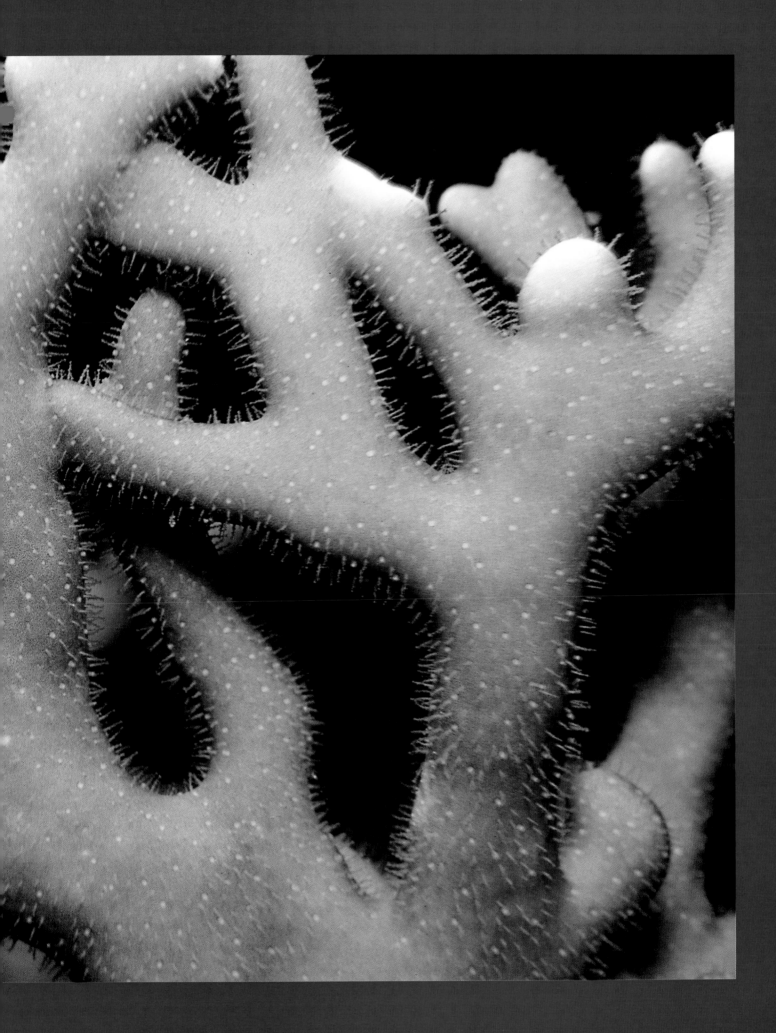

Sea fans (**Left**) and gorgonians (**Top and Above**) are
the most colourful of the corals.

Both sea fans and gorgonians have a branching mesh-like structure covered in polyps (**Above, Above Right and Right**). The mesh size is determined by the current flow. If a floppy fan with a wide mesh is moved to a more exposed place with stronger currents, it will develop a smaller mesh for greater rigidity. Most sea fans grow at right angles to the current making it almost impossible for plankton to pass through the fan without being caught. Gorgonian corals are found under ledges or attached to cave roofs, possibly to avoid too much light. Their colour comes from the hard, but flexible material called gorgonian in their surface tissue, which overlay rods of brown tissue within. These corals are found in the deeper reefs so that their flexible skeletons can gently move in the current, rather than be buffeted by the waves (**Overleaf**).

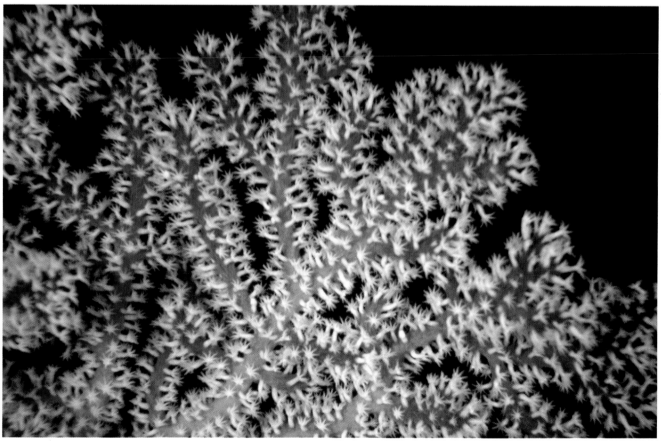

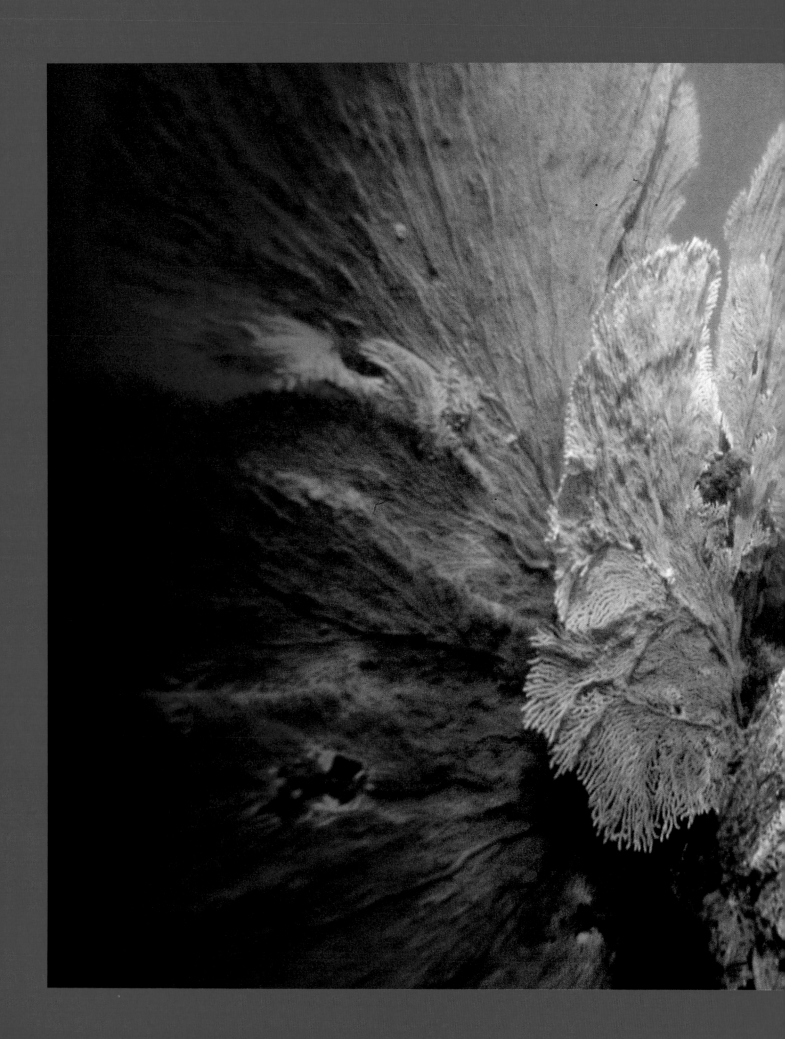

Soft corals lack a limestone skeleton (**Above, Left and Right**). Their flexible support is provided by hard spines called sclerites, which are embedded in their tissues. Some white soft corals lack zooxanthellae, so their polyps are responsible for trapping plankton in the water. These soft corals have skeletons that look white within a transparent outer covering (**Overleaf**).

Coral polyps are withdrawn during the day when they may be grazed by fish. They come out at night (**Above, Left and Right**), when planktonic animals from the sandy floor, crevices and deeper water move over the reef. Hence it is at night that the corals reveal their true colours. Many smaller animals can be found hiding amongst the coral polyps including tiny porcelain crabs (**Overleaf**).

Previous Page: As well as the corals there are other members of the sea anemone family on the coral reef, such as the attractive jewel anemones.

The larger corals found in deeper water need sheltered conditions. Here lion fish swim between large fan corals growing at the base of a coral wall (**Above**). The golden swift coral (**Left**) is a brightly coloured soft coral found growing in the Solomon Islands. There is serious competition for space between the corals of the reef and faster-growing corals will try to overgrow a slower one. Others undertake direct action by stinging their neighbours.

The larger plate, table (**Above**) and cone (**Left**) corals are found in sheltered areas or in moderately deep water below zones of wave action. Brain corals (**Overleaf**), grow up to two metres (6.5 feet) in diameter but never too deep as they need light for their zooxanthellae which photosynthesise. They grow extremely slowly so a brain coral one metre (three feet) across could easily be several thousand years old.

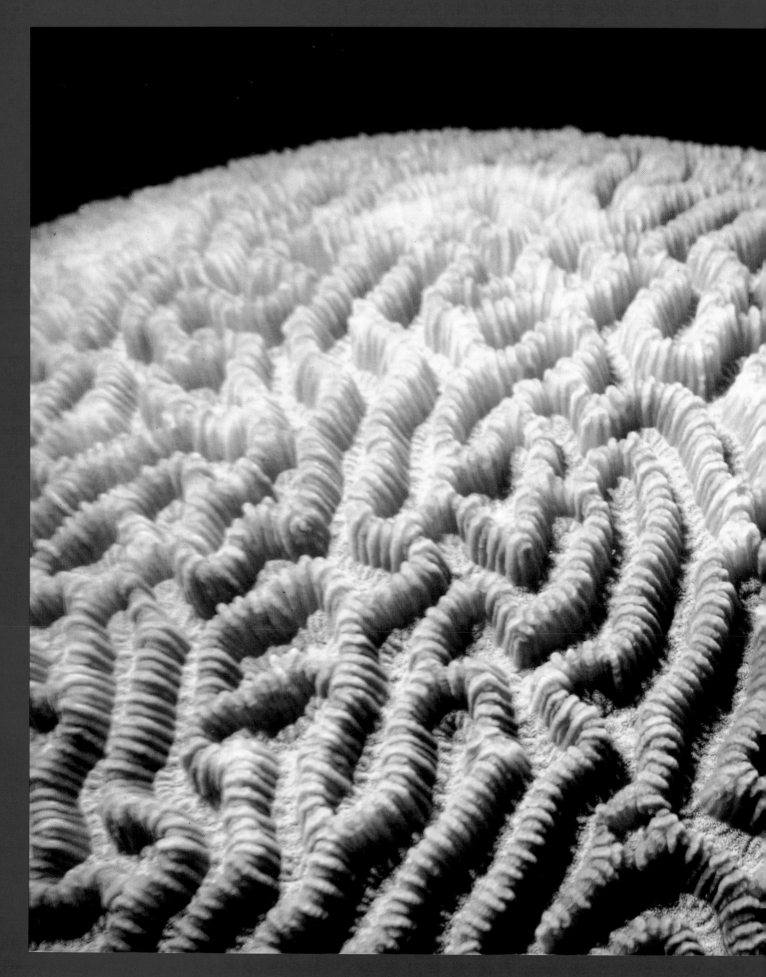

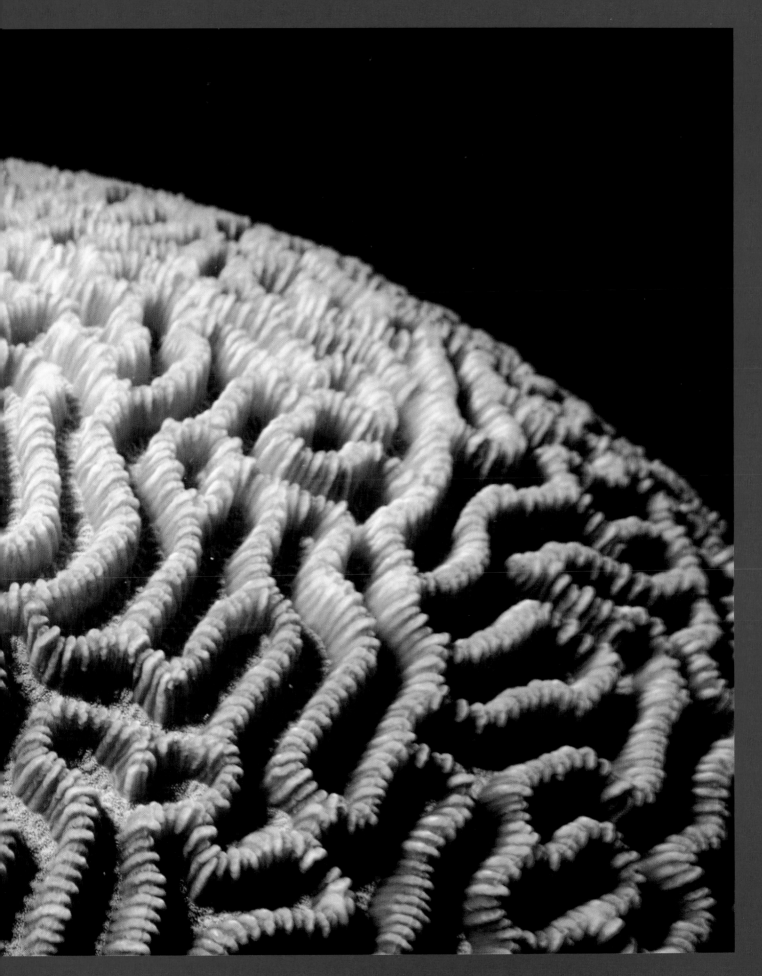

284

Above: This weird, hand-like outgrowth of a sponge could easily be nick-named "ET".

Above Left: There are crevices and caverns at the base of the coral slopes where the action of the sea has eroded the coral. At night, fish take shelter in these caves.

Overleaf: Some corals have a strange appearance such as this bubble gum coral.

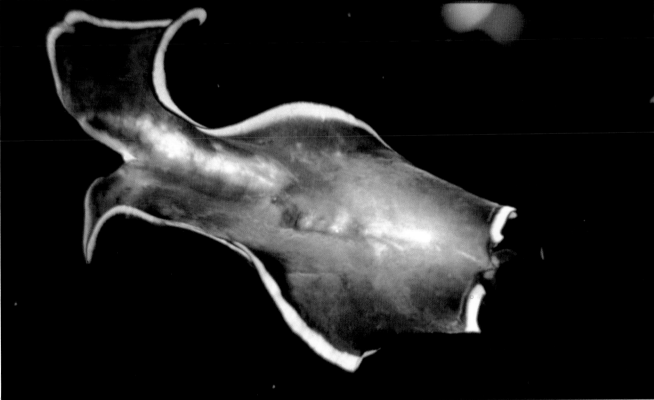

Above Left: The corals provide a base to which other animals can become attached, such as this large feather star. Animals are also found in all the nooks and crannies between the corals.

Left: Fan worms extend a ring of sticky tentacles into the current to filter out plankton.

Top: The bearded fire worm is an active hunter on the coral reef. Its body is covered in tiny white hair-like projections which act like paddles, moving the worm forwards. They inject poison on contact, hence their name.

Above: A small flat worm swims across the reef.

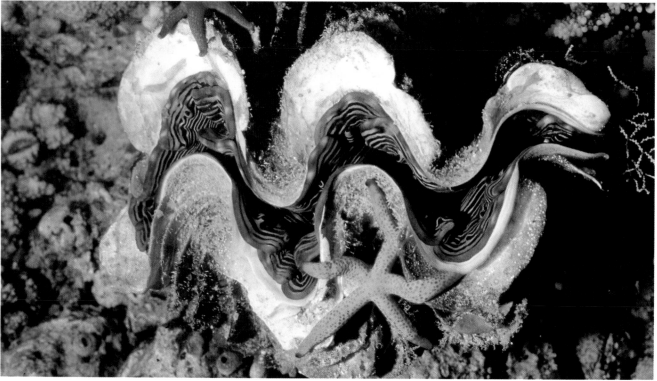

Clams and oysters display delicate mantles that form a siphoning mechanism for feeding on plankton in the seawater. They have simple eyes that are sensitive to light. If they detect nearby movement they quickly clamp their shells tightly together for protection (**Above Left**). They are attacked by starfish (**Left**). The starfish digests its prey by pouring onto it the contents of its stomach, and then sucking up the soluble products of the digestion. The bright colours of the mantle of the giant clam (**Above**) are a result of the diffraction of light by sub-microscopic layers of a crystalline non-coloured pigment. Each clam has a unique pattern of colours—a form of fingerprint—and some giant clams are more than 100 years old, weighing up to 260 kilograms (575 pounds). The clams open to the sun during the day so that the mutualistic algae in their mantle can photosynthesise and provide the clam with some of its food.

This spindle cowrie (**Far Right**) is feeding on a gorgonian coral. Cowries are active at night (**Above**). The old shells may be the home of hermit crabs (**Right**) but as the crab grows larger it has to find a larger shell. The allied cowrie (**Overleaf**) is camouflaged to look like the coral on which it hides and feeds so that it is not spotted by predators.

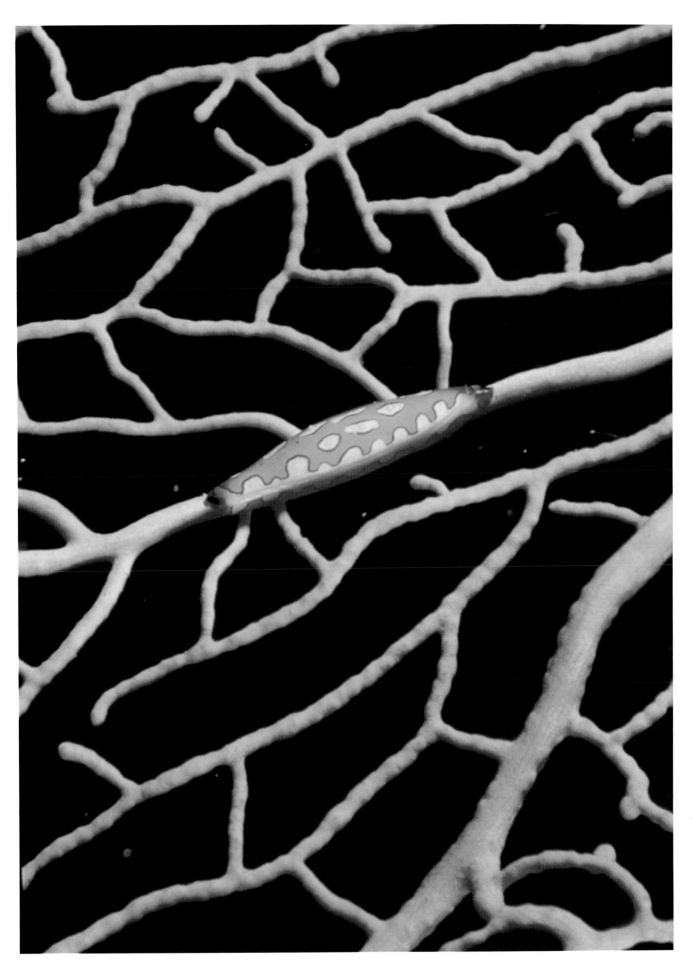

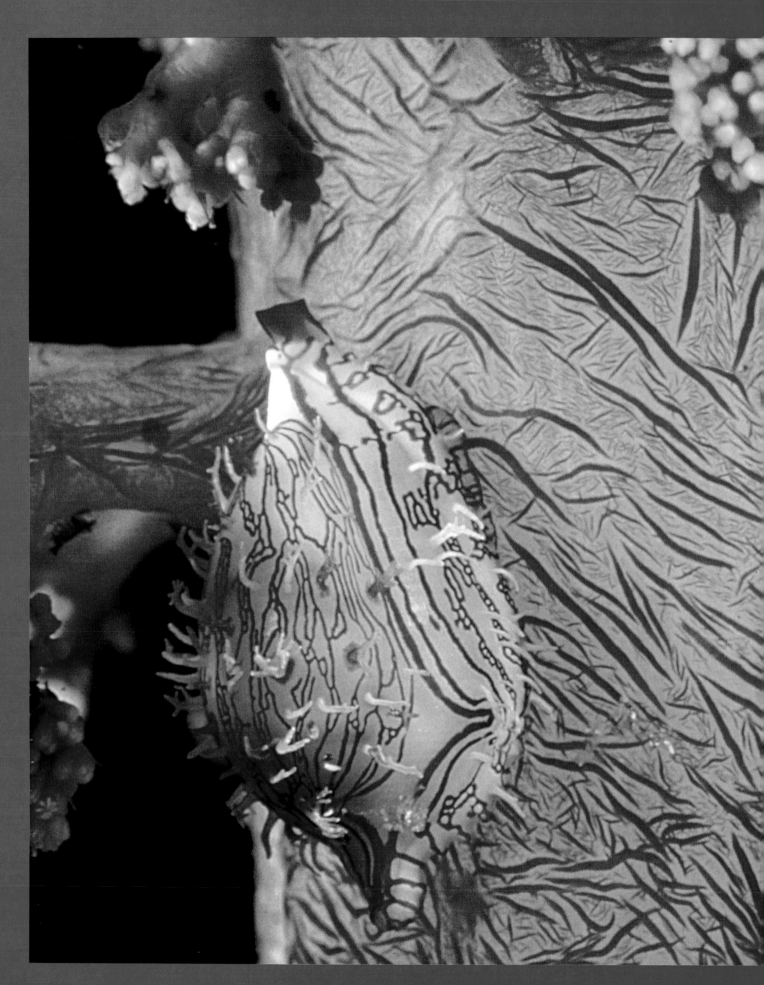

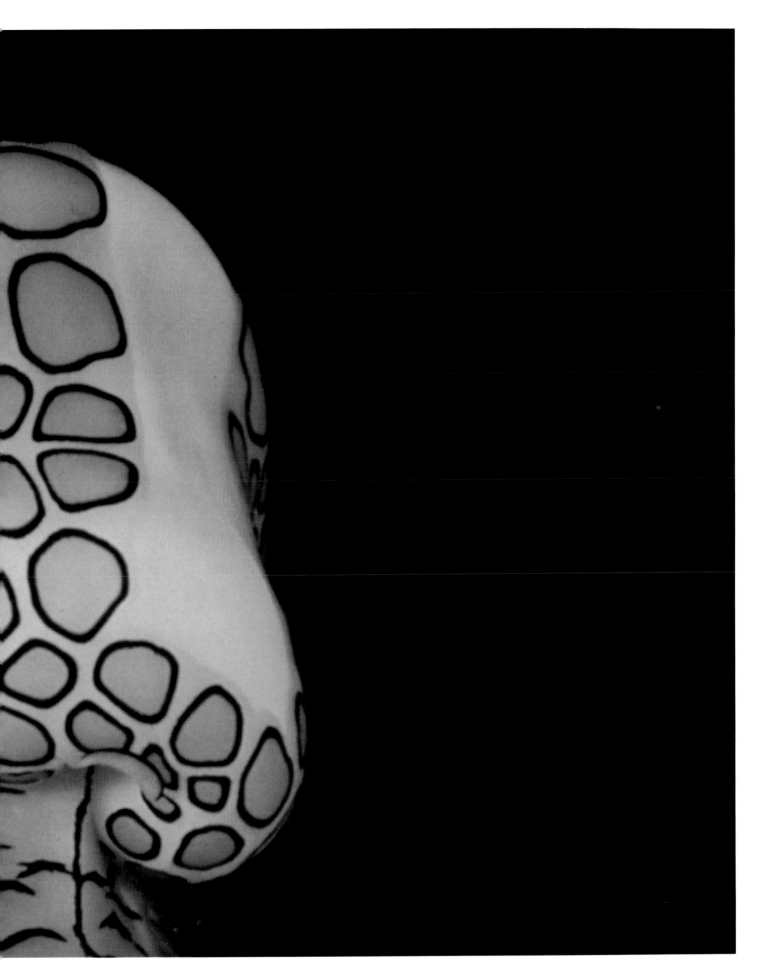

Previous Page: The brightly coloured flamingo tongue snail feeds on soft corals.

The cat's eye turban snail (**Left**) is named after the appearance of the underside of the snail when its foot is fully retracted into its shell (**Overleaf**).

Top: The scorpion conch has a shell with many long spines.

Above: The harp shell is amongst the largest gastropods. Its huge body is too large to be able to retract inside its shell.

Right: Top shells have a simple spiral shell with a pointed top.

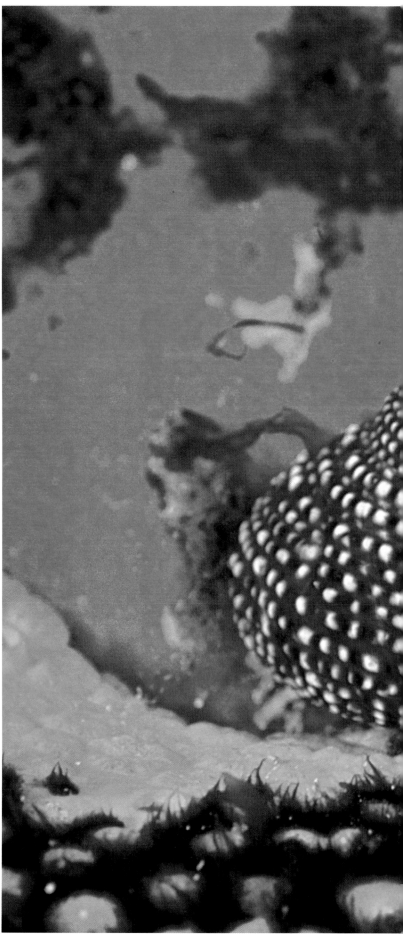

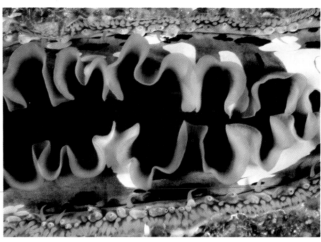

The thorny oyster (**Left**) is a bivalve and is related to the clams and mussels. When its shells part it is possible to see the many brightly coloured folds of its mantle (**Above**).

The pattern of skin colours of a cuttlefish reflects its moods (**Left**).
Cuttlefish can change the colour of their skin in order to communicate
with others of their species. During courtship, the patterns change rapidly
from red through to cream and back again (**Above**).

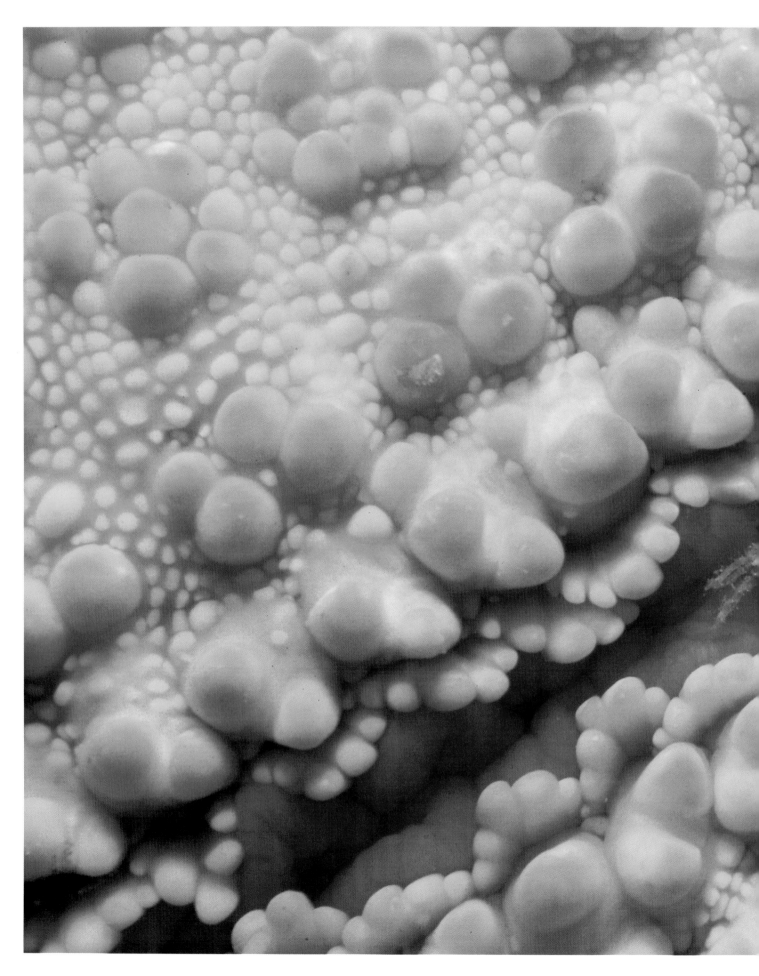

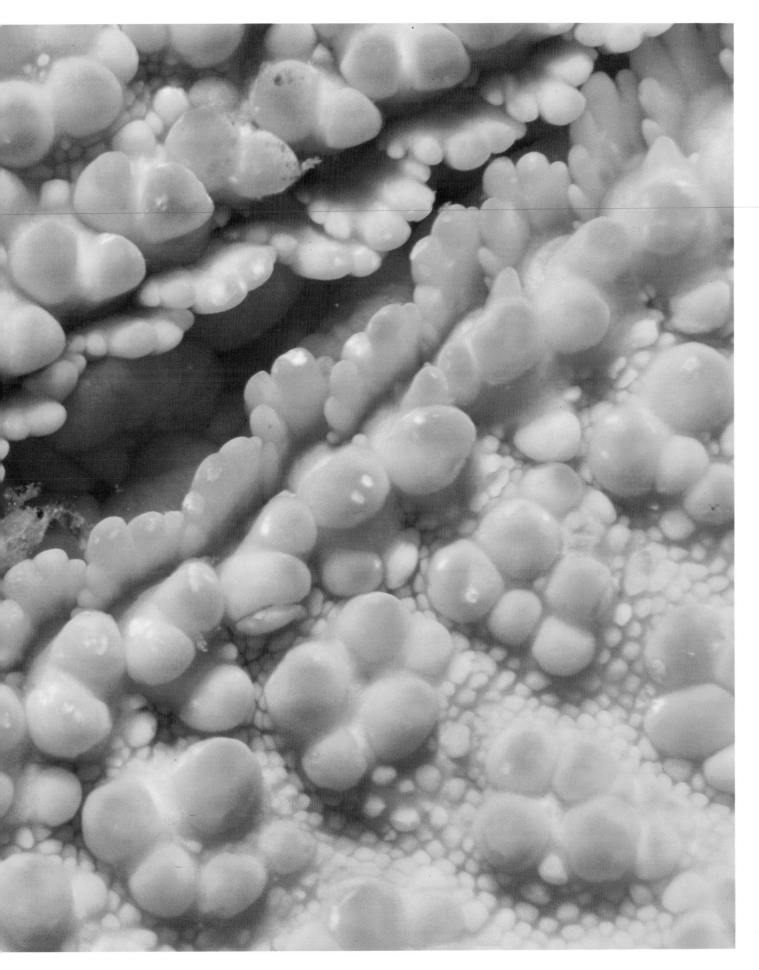

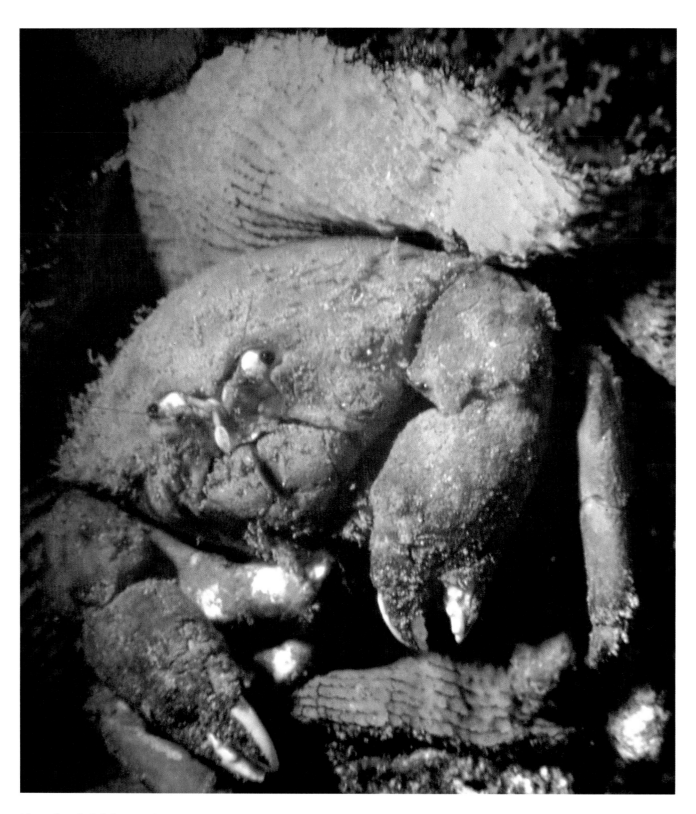

Minuscule crabs hide between the spiny plates of cushion stars (**Previous Page**) and on the rough skin of sea cucumbers (**Overleaf**). The sponge crab (**Above**) carries around a protective covering of sponge held in place by the last two pairs of walking legs. Sponges are distasteful to most reef animals so by carrying the sponge the crab gains some protection. The decorator crab covers itself in tiny bits of coral for camouflage (**Above Right**). Porcelain crabs live amongst anemones for protection, feeding by filtering plankton from the surrounding water (**Right**).

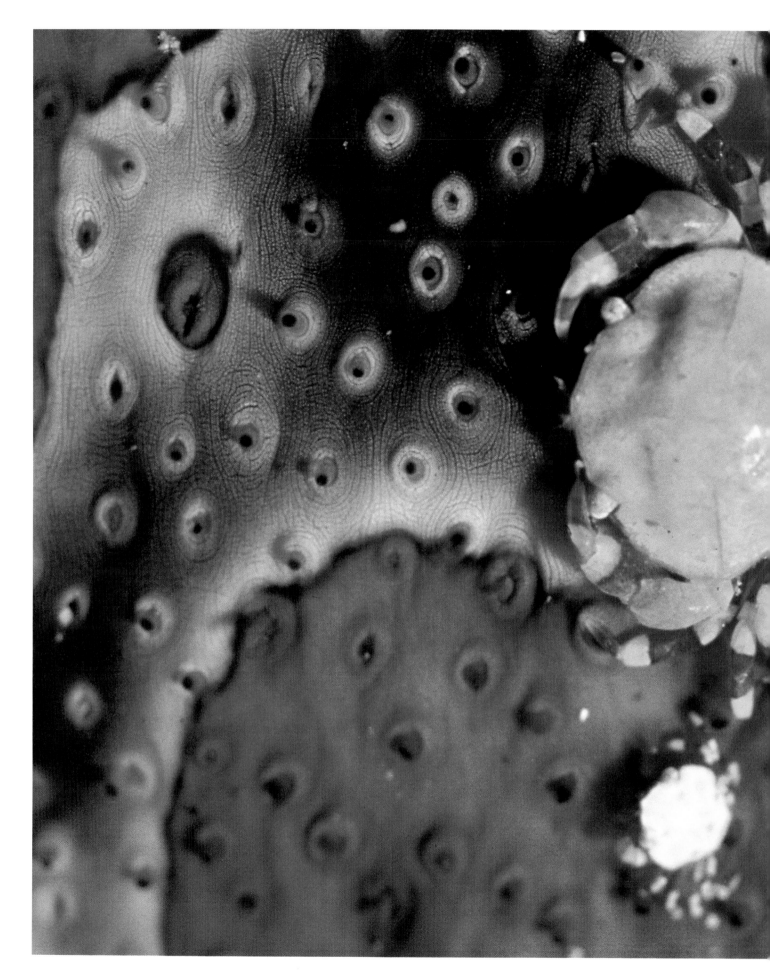

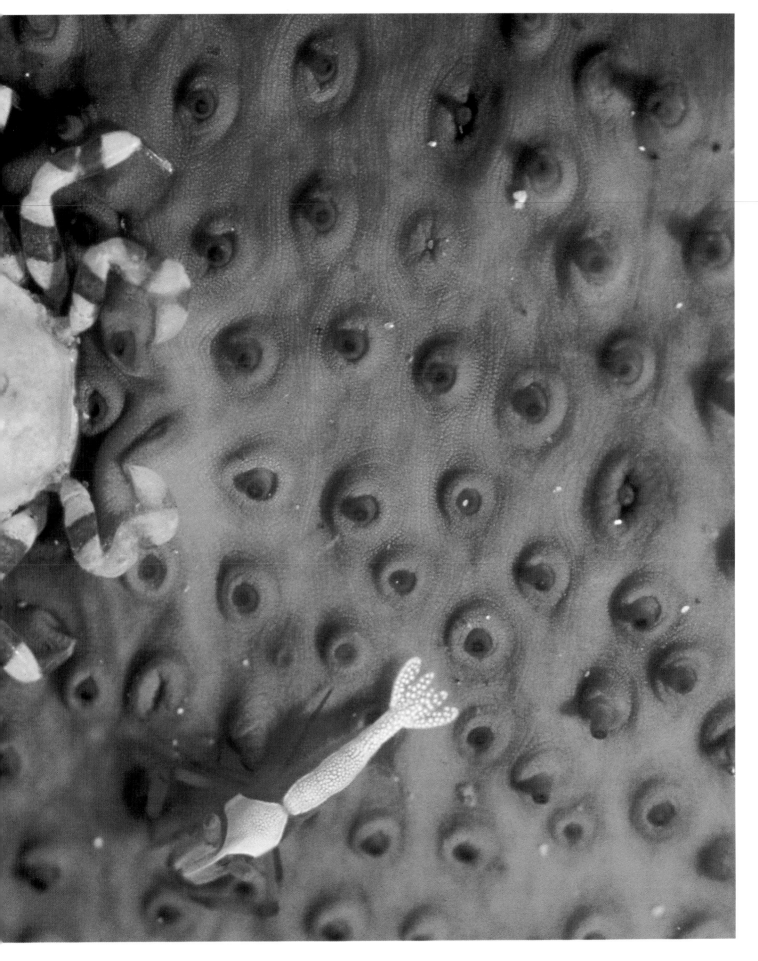

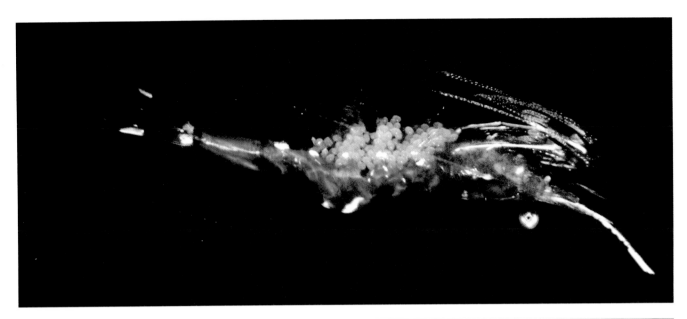

A typical shrimp has a flattened transparent body with a fan-like tail (**Top**). The banded cleaner shrimp (**Above**) is visited by fish. It stands in a conspicuous place so the fish can see it. The shrimp then nibbles over the surface of the fish, feeding on its mucus and any parasites. The mantis shrimp (**Above Right**) can grow to around 15 centimetres (six inches), and resembles a heavily armoured caterpillar. The head is densely spined and two club-like claws are ready to pound prey. The mantis shrimp is an extremely efficient predator, taking shrimps, crabs and fishes. The spiny lobster (**Right**) is an important predator and scavenger found under ledges or in caves with only their long antennae sticking out. They use sound to warn other lobsters, and even human divers, away from their shelter on the reef. The lobsters produce a grating or buzzing noise by rubbing hard pads at the base of the antennae against special ridges on their head—an activity called stridulation.

Above: Lobsters, such as this painted lobster, have a pair of appendages on almost every segment. Those on the abdomen are simple swimming flaps but those on the head and thorax are more specialised for feeding and moving.

Above Right: Most tropical sea cucumbers produce toxins called holothurin. When threatened, some sea cucumbers point their anus at their attacker and discharge long white tubules. As these are extruded, they split, releasing their poisons and a strongly adhesive substance that will immobilise a small fish or crab—a bit like a very strong glue.

Right: Feather stars, or crinoids may have anything from five to 200 arms, which are held into the current to trap plankton. They select a site where there is a good flow of water and some adopt a bushy multi-layered feeding position, which helps if the current is variable.

Above: The sea urchin is protected by long spines that can be venomous. They have long tube feet, which, together with the spines, move the animal over the coral.

Above Left: A spawning starfish in the Galapagos. The free-swimming starfish larvae will form part of the zooplankton before settling back on the reef to become an adult.

Left: The brittlestar is related to the starfish but its pentagonal shaped body is much smaller that a starfish's and the legs are long and thin.

Overleaf: The tunicate is characterised by a protective covering called a tunic containing cellulose, which is rare in animals. There are two openings, one for the entry of water into the filtering apparatus and one for the exit.

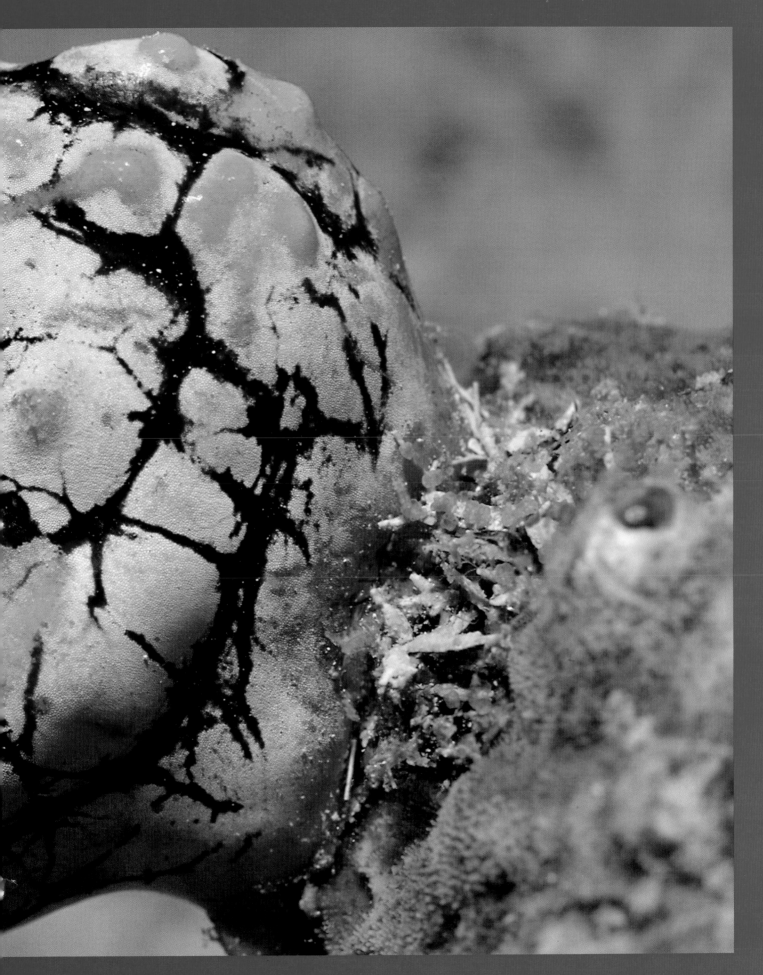

Above: These wavy structures are the eggs of the nudibranch known as the Spanish dancer.

Left: The Christmas-tree worms bear little resemblance to common worms. These colorful creatures make a home by burrowing into a coral, then display their Christmas-tree shaped heads to filter plankton from the surrounding water. Sensitive to light and pressure, any change causes them to flick down inside their holes.

Right: More species of fish are found on the coral reef than in any other marine habitat.

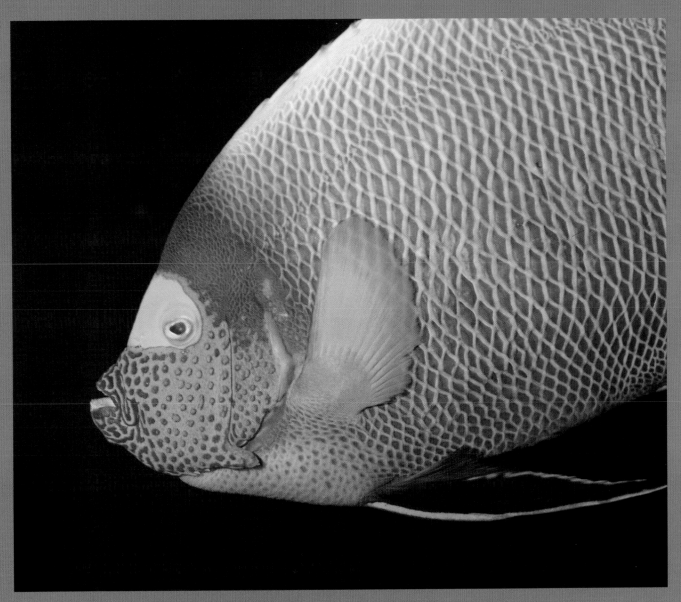

Angel fish, such as the blue-faced angel fish (**Above**) and the Emperor angelfish (**Above Left**) are large colourful fish, found as singles or in pairs in areas of rich coral growth and at depths up to 30 metres (100 feet).

The bright colours of the reef fish are often used for communication. They have to be able to stand out to grab the attention of potential mates or to posture in front of would-be competitors. Seen here are the distinctive markings of the black and white grouper (**Top**), yellow tang (**Left**) and the coral grouper (**Right**).

Above: The powder blue surgeon fish is so named because of the sharp, curved, blade-like spines on either side of its tail. A flick of the tail will injure a predator.

The members of the sweetlips family are favourites of the angler, but they are difficult to catch, hence they are also known as the tricky angler fish!

The oriental (**Previous Page and Right**) and harlequin sweetlips (**Below Left and Overleaf**) feed actively at dusk. The appearance of the juvenile sweetlips (**Left**) is very different from that of the adult. This oblique-banded sweetlips is having surface parasites removed by a visiting cleaner wrasse (**Above**).

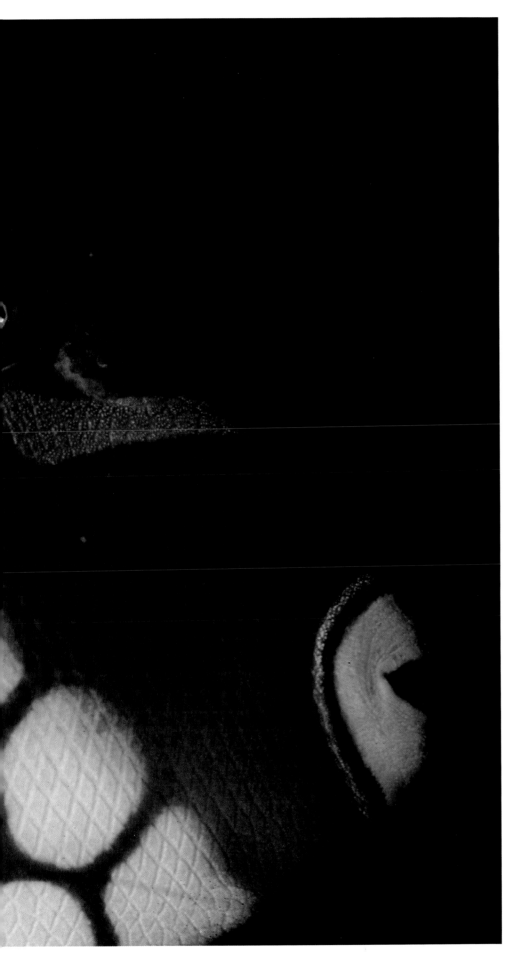

The triggerfish has an unusual method of defence. The front of its dorsal fin has evolved into a sharp spine that will erect and lock into place. If chased into a hole in the reef it can use this sharp spine to wedge itself into the hole. A predator might be able to nibble at its tail, but would not be able to pull the worthwhile meal of the fleshy part of the fish out. Triggerfish also have an aggressive temper and sharp incisors. Given an opportunity they will often turn on an attacker and inflict a nasty wound. When guarding their eggs, some species have even been known to attack divers. Seen here are the blue triggerfish (**Top**), redtail triggerfish (**Above**), and orange-lined triggerfish (**Overleaf**).

Left: The parrot fish starts life as a drab female and has at least one breeding season before turning into a male. The males keep a harem of females and if anything happens to the male, a female changes sex to take his place.

Previous Page: The squirrelfish gets protection from a poisonous spine located on its cheek.

Above: The masked butterfly fish feeds on coral polyps covering a wide area, especially at the coral edge.

Left: Streamlined barracuda are normally seen in shoals when young, like these. They are active predators, hunting small fish.

Many of the reef fish have a flattened body, like the moorish idol (**Right**) and the batfish (**Above Right**).

341

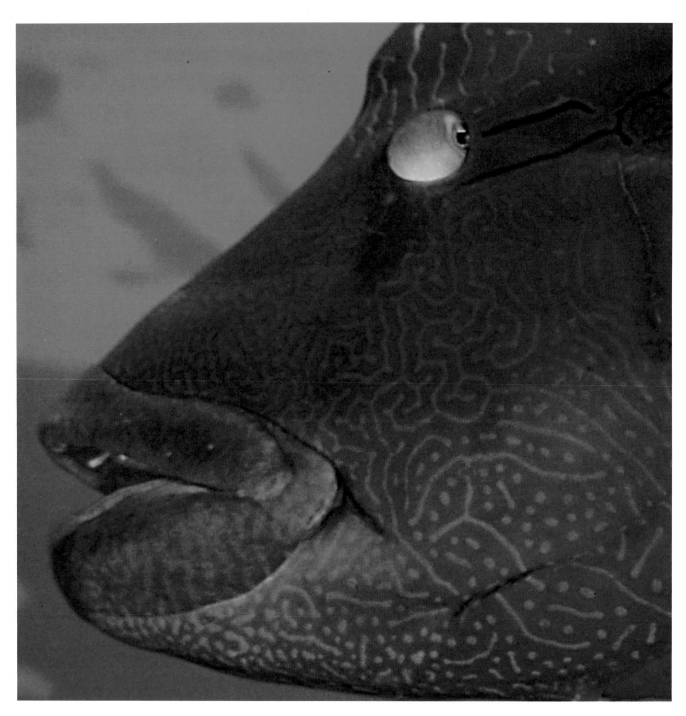

One of the largest fish of the reef is the huge napoleon wrasse (**Above, Right and Overleaf**). It grows up to two metres (6.5 feet) in length and lives for many years. Its huge size can be clearly seen as it swims amongst a shoal of blue-striped snapper (**Right**).

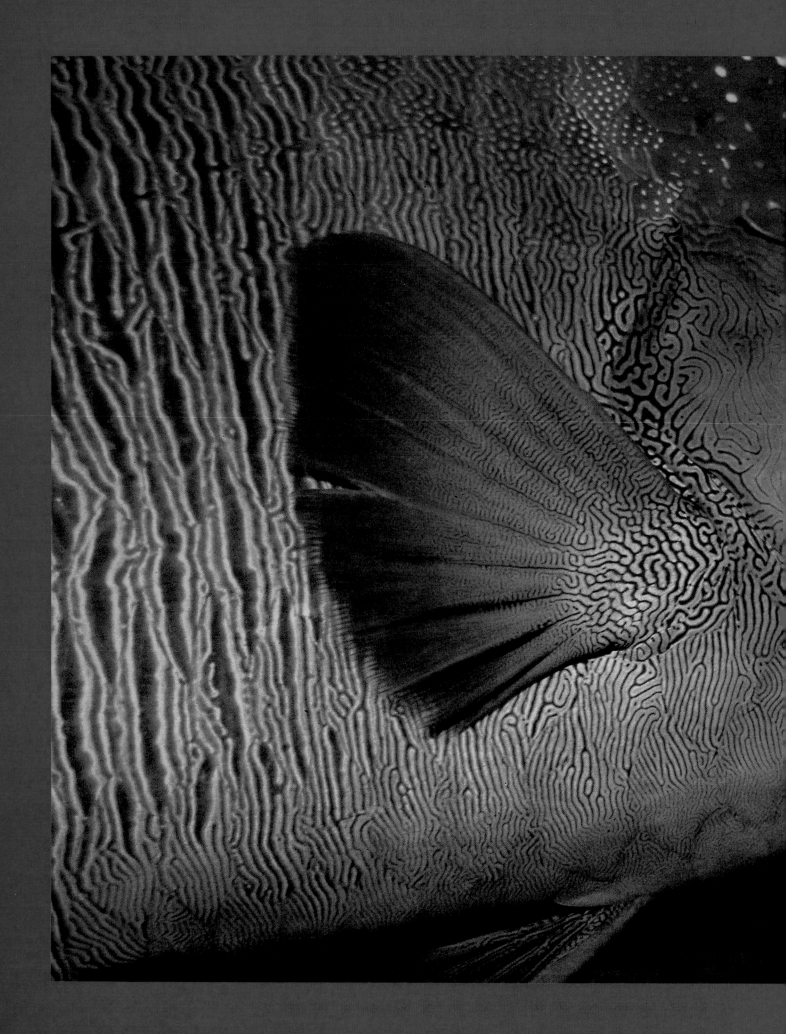

Bright shoals of yellow-backed (**Above**) and yellow top (**Left**) fusiliers patrol the reef during the day, but retreat into the coral at night. Then they change colour becoming drab blue with a mottled appearance and rest motionless in the dark amongst the coral.

Far Left: The puffer fish has a strong beak to feed on hard-shelled animals such as sea urchins, molluscs and crabs. Puffer fish are dangerous to eat as they have a poison in their liver and sex organs.

Shoals of black spotted grunt (**Above**), cromis (**Right**), jacks (**Left**) and soldier fish (**Overleaf**) move as one across the coral reef. Many fish live in shoals for protection.

Right: Many shoals consist of more that one species; here blue striped grunts and margates swim together for added protection.

Overleaf: A shoal of harlequin sweetlips swim over a coral reef off the coast of Malaysia.

Many fish pick up parasites such as lice, like this blotched hawkfish (**Top Left**). The cleaner wrasse is an important fish on the reef as its job is to pick off any parasites attached to other fish, as it is doing to this coral rockcod (**Top**).

There are many different types of rockcod on the reef, including the coral rockcod (**Above Left**), the peacock rockcod (**Above**) and the white-lined rockcod (**Far Left**).

Left: The mottled appearance of the leather bass is good camouflage against the reef.

357

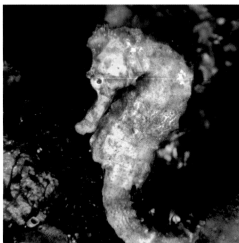

Often it is impossible to believe that some of the strange shapes on the reef are actually fish. The devil stinger (**Top**), the longsnout seahorse (**Above**) and the crab-eyed goby (**Below**) are all very well camouflaged and very wierd looking fish!

Right: This pair of blotched hawkfish have almost perfect camouflage against the pink colours of the soft coral.

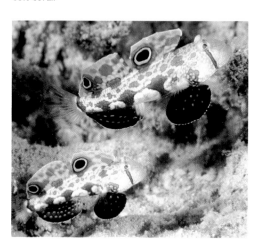

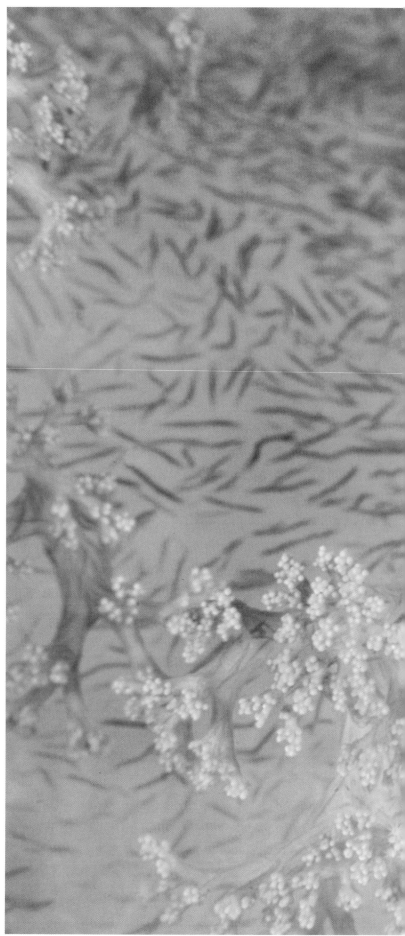

Glassfish (**Above Left**) are semi-transparent with silvery and gold reflective patches. Their size varies from a few millimetres to a few centimetres, but in any particular shoal each fish is of similar size, from the same generation. The reason is obvious; its larger cousins would eat any smaller fish joining such a shoal. As the shoal writhes within its shaded location, light catches the reflective patches on the fish to give a twinkling effect making it difficult to focus on any individual glassfish within the shoal. The overall benefit of this behaviour is to confuse predators. Reef dwelling predators such as the black-spot snapper (**Left**) and humpback grouper (**Above**) will slowly approach shoals of glassfish, then on reaching a critical distance, dart through the shoal to snap at individual fish. It is from this hunting behaviour, that snappers acquired their common name. As the predator passes through the shoal of glassfish, a tunnel of clear water appears as the prey surge out of the way. Emerging from the other side, the predator will have already swallowed any fish unlucky enough to be caught.

This Page: Younger barracudas often form enormous spiralling shoals. When it comes to hunting, often the first sign of a barracuda making an attack will be its tail disappearing away after it has snatched an unlucky fish.

A lionfish hangs virtually motionless in the water (**Below**). Its spiky outline and striped colouration do not blend completely into the background, but can confuse a victim as to which direction it is facing. By moving slowly, the lionfish allows the target to become used to its presence, until a sudden movement and a big gulp and the victim is swallowed whole (**Right**).

Above: Two lizard fish lie motionless on the coral.

Right: A tiny blenny sticks its head out of its hiding place.

Tiny gobies hide among the corals. The striped goby (**Above**) and many-host goby (**Right**) are dwarfed by the corals. Two masted shrimp gobies lie beside shrimps on the seabed (**Below**).

Tiny reef animals such as ostracods (**Left**)—a type of crustacean—and sea spiders are found living among the larger animals on the reef, such as sea squirts (**Above**).

The coral cat shark is very similar in appearance to the dogfish found in temperate waters. This small member of the shark family is active at night.

The irregular outlines of the ghost pipefish (**Above Left**), pipe fish (**Left**) and seahorse (**Above**) help conceal their bodies against their background.

Sting rays gather in groups or schools on sandy beaches at high tide (**Top**). They can be inquisitive animals and here they are quite happy being fed and handled by divers (**Above**).

The snake eel (**Above Left**) lies concealed in a hole in the ground during the day, while its relative the leopard moray eel (**Left**) hides in a crevice.

The outer slopes of the coral are rich pickings for plankton feeders, as the ocean brings in loads of plankton. Many fish swim in shoals at the coral edge, but have to be wary when swimming too far from the protection of the reef, as they can be caught by the sand tiger shark (**Top**). The leopard shark is another predator of the reef (**Above**).

Above Right: Venomous sea snakes swim in the shallows of the reef.

Right: At night many fish seek shelter in the coral. Their bright colours becoming far more subdued. This sleeping parrot fish is much paler at night compared with its daytime colours.

MASTERS OF DISGUISE

There are many predators patrolling the reef looking for food, so fish have to camouflage themselves to look less conspicuous. Predators too take to concealing themselves so that they can ambush unsuspecting prey as it passes by.

Above and Opposite: The frogfish is also called the angler fish and it is an ambusher. It is also the master of camouflage as it can change the colour of its skin to match its surroundings, as can be seen in these photographs. The lack of a defined body shape also makes the frogfish difficult to spot. Its first dorsal spine is modified as a unique fishing pole with a lure and when it is wiggled a fish may be attracted to it and will then be sucked in whole.

Left: The colourings of this pygmy seahorse allow the tiny fish to blend in with its surroundings and prevent it from being eaten. It also has knobs or protrusions on the surface of its body that resemble those of the coral in which it lives, rendering it inconspicuous.

Below, Far Left: Another camouflaged bottom-dwelling predator is the Indian flathead. Unlike the scorpionfish and stonefish it does not have poisonous spines but it does have a large mouth full of teeth and when threatened it does not try to escape but faces its predator and attacks.

Below Left, Centre: A scorpionfish has a ragged appearance, with fins and body surrounded by delicate tassels. Its skin is green, with a typically brown and red mottled coloration to blend in with the coral on which it rests. Its feeding strategy is to lie in wait, merging into its surroundings, until an unsuspecting meal swims within gulping distance.

Below Left: In order to be camouflaged an animal does not have to resemble the reef or rocks. The lacy scorpionfish is quite different from other scorpionfish as it has permanently extended dorsal and pectoral fins that are broken by areas of transparent skin. Nestling on an exposed coral head this camouflage is designed to make the fish look like an inoffensive feather star.

Below: The leaf fish does not try to blend with its surroundings but instead it looks like a dead leaf, with its chin barbel resembling the leaf stalk. It escapes predation by tricking its predators into assuming it is dead plant material and therefore unpalatable to them.

Bottom: The snake eel disguises itself by mimicking the shape and colourings of the venomous sea snake. Thus it has the advantage of looking poisonous without having to go to the trouble of manufacturing a poison.

CORAL ABSTRACTS

Macro photography introduces us to a world within the world of the coral reef. Many divers who are lucky enough to see the amazing diversity of organisms living on the coral reef are concerned with capturing on film new species they haven't seen before. Also the temptation is always there to look for the giants of the reef like the sharks. Here we can appreciate sights most people never see; shapes, colours, shadows and patterns to delight the eye of the beholder.

Above: Close up of an anemone showing the beautiful purple coloration and the tentacles retracted to form a button shape.

Above Right: Mantle of a clam showing the region under its shell where exchange of gases occurs with the surrounding water.

Right: Close up of a crown of thorns starfish.

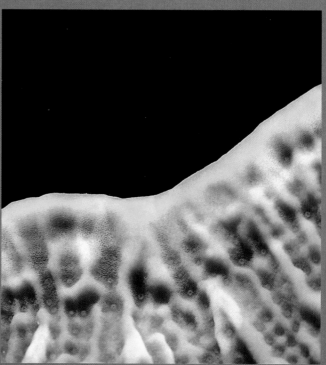

Left: Coral colonies in Indonesia showing a calcareous skeleton.

Below, Far Left: Skin detail of the emperor angelfish showing the gaudy colours, which blend amazingly well as they flutter among sea fans or nibble on bright-hued sponges.

Below Left: The pink-tinged edge of a coral highlighted against the inky blackness.

Right: The surface detail of a crayfish carapace, which is the hard external skeleton covering the body of this crustacean.

Below: Close up of soft coral, showing the individual red polyps and the hard sclerites that are embedded in the polyps' tissues.

Below Right: A sponge, showing the opening where the creature takes in water.

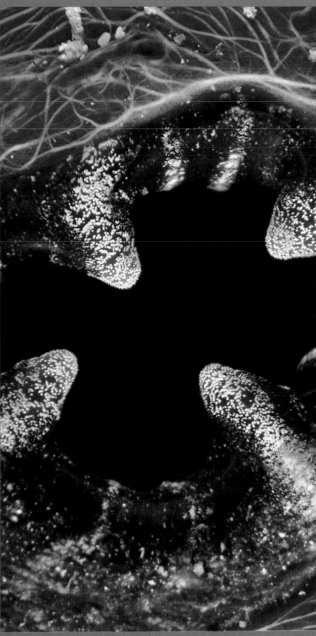

Above: A close up of a crinoid showing the delicate feather pattern that gives it the name feather star.

This Page: The detail of the soft folds that make up the mouth of an anemone.

Overleaf: A close up of a cushion seastar, showing the mouth, five radiating grooves and the suction cups of the tube feet.

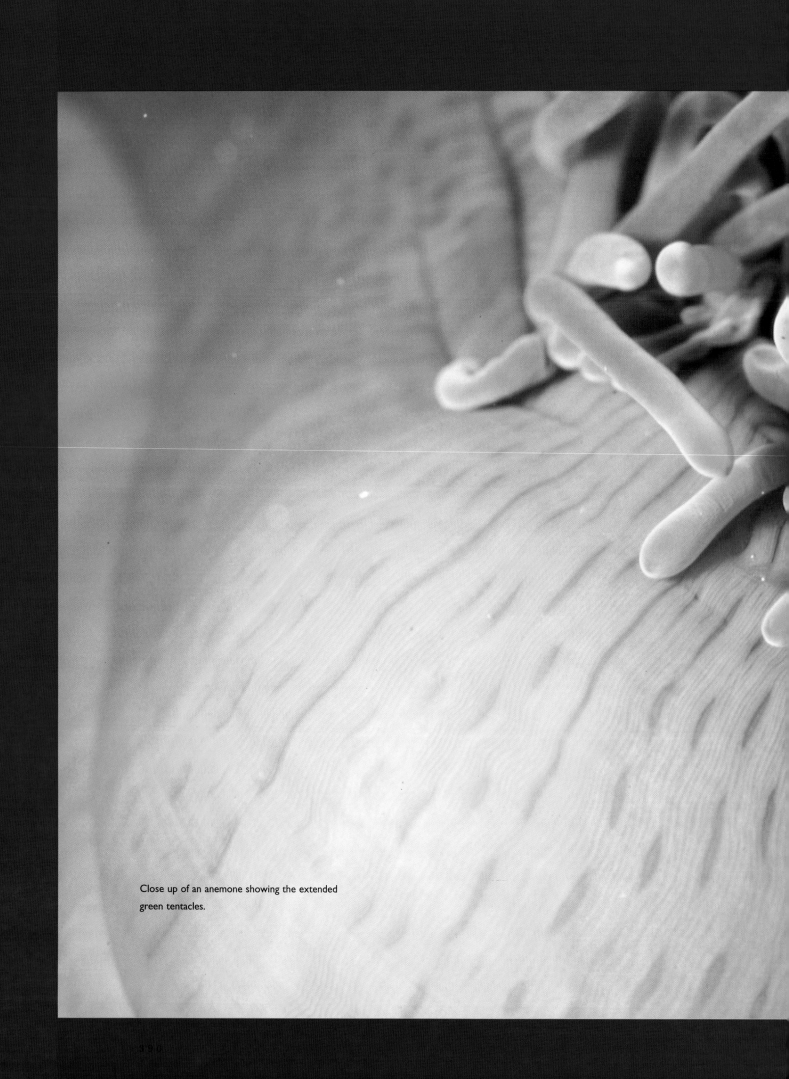

Close up of an anemone showing the extended
green tentacles.

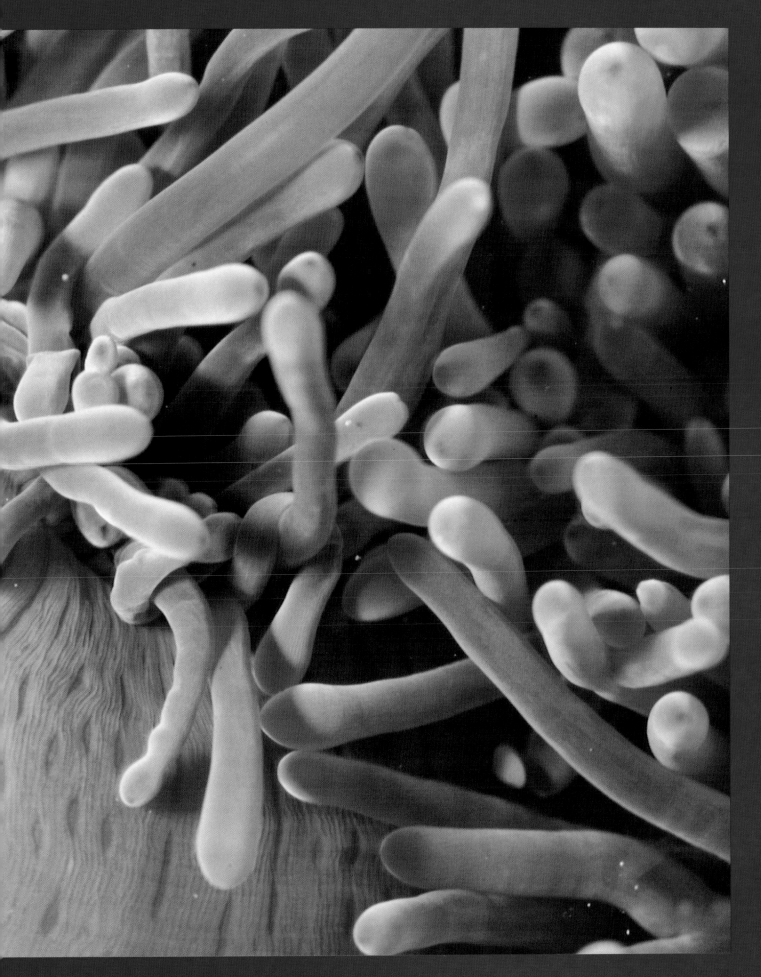

THE GREAT BARRIER REEF

Stretching more than 2,000 kilometres (1,250 miles) along the coast of Australia, the Great Barrier Reef is the most famous and largest scale example of barrier reef. It is also the largest structure on the planet made by living organisms and it can even be seen from space. The reef consists of more than 3,000 individual reefs including 760 fringing reefs and 300 coral cays (sand islands). There are just over 600 continental islands that were once part of the mainland, but have been cut off by the rising sea level. The species list is impressive— 1,500 fish, 400 corals, 4,000 molluscs, 500 seaweeds, 215 birds, 16 sea snakes and 6 sea turtles. Not surprisingly, it was given national park status in 1975. In 1981, it was made a world heritage site in recognition of its global importance. Today, it is the world's largest protected area.

Management of the national park is critical as there are several conflicting interests—fishing, tourism and conservation. The rich fisheries of the seas around the reef support a large fishing industry, especially shrimp boats. Over the last few years, visitor numbers have soared and now there are almost 2,500,000 visitors each year who visit the park to fish, dive, snorkel, water ski, sightsee and reef walk. Some areas have already reached maximum capacity and there is pollution from sewage, chemicals, fertilisers and pesticides. Just jumping into the ocean wearing sun-block contaminates the water. Animals that are sensitive to disturbance and pollution are decreasing in some of the busiest areas, for example the number of dugong near Dunk Island have halved in recent years.

Above: Erskine Island is a coral cay. It is located in Zone B where there is no collecting or fishing and limited camping.

Right: Wistari Reef is a completely enclosed lagoon with a mosaic of patch reefs and shallows, each like a mini reef in its own right.

Below Right: The Great Barrier Reef extends along the coast of Queensland, Australia.

The park has been divided into zones with different types of permissible activities. The General Use Zone is where most of the tourist activities take place such as fishing, water sports, diving and camping. In Zone A, there is camping but no collecting or fishing and limited crabbing. In Zone B, the camping is limited to a certain number of days per year. There is a scientific research zone with limited access by permit for research purposes. And finally, a preservation zone with no access. Preservation zones make up less than four per cent of the total area, but they cover most of the important reefs and turtle breeding beaches.

The Great Barrier Reef is the world's greatest natural wonder—let's hope it stays that way.

Top Left: Deep sea fishing boats anchored at the edge of the reef where there is plenty of good fishing.

Above Left: The crown of thorns starfish has devastated large areas of coral. However, biologists now believe that it can maintain coral diversity. It tends to feed on the faster-growing corals, which if left unchecked would dominate the reef.

Left and Above: Green Island is an idyllic island, but each day hundreds of day trippers arrive from the mainland to swim, reef-walk, snorkel and dive. The latest fast catamarans mean that more than 80 per cent of the park is now accessible to day trippers.

Above: A fringing reef at low tide showing the coral formations.

Above: A bright blue starfish stands out against the corals.

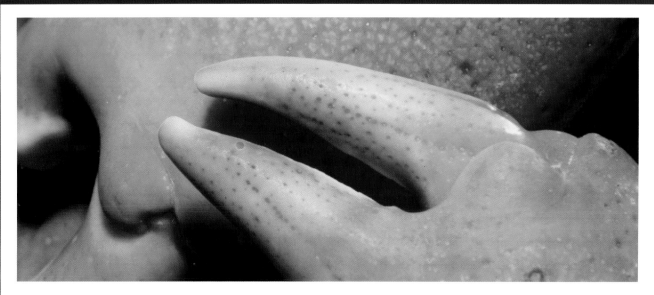

WEAPONS OF WAR

It is out-and-out war in the ocean, especially on the coral reef, where the intense predator pressure and competition for space amongst sedentary organisms has led to a wide variety of defensive adaptations in reef animals.

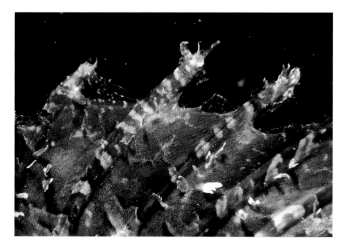

Above: Crabs are among the largest of the crustaceans and their most characteristic feature is their powerful, frightening looking pincers, which are primarily used for feeding. However, the claws also function as sex symbols, males often having one claw much enlarged and strongly marked, which can be brandished as a threat to a rival, as well as to impress a mate. They also use their claws defensively, standing their ground when threatened rather than retreating.

Left: The stonefish has 13 dorsal spines which are sharp, sturdy and grooved on each side. There are poison sacs on each spine that inject such a painful venom that the agony can drive a victim insane. Madness is soon followed by numbness and death.

Right: The moray eel is a largely nocturnal fish that hides in crevices and under coral ledges with its vicious looking head protruding, always on the look out for food passing by. Its narrow muscular jaws can drive its fang-like teeth deeply into anything they grasp. The bite itself is not toxic though and moray eels are generally harmless to humans unless provoked.

Above Left: Cone shells are single-shelled molluscs that have adapted their ribbon-shaped tongue, covered with rasping teeth, into a stalked radula, modified into a sort of gun. They hide in the rubble of the seabed by day and cunningly extend the radula towards their prey, which may be a worm or even a fish, and then they discharge a tiny, glassy harpoon from the end. The victim is caught by this and whilst it struggles the cone shell injects a virulent poison that can kill a fish instantly. They then haul the prey back to their shell and slowly digest it. The venom of a cone shell can even be lethal to humans.

Left: The file or flame shell is a venomous bivalve mollusc.

Above: This venomous sea urchin is an echinoderm like the starfish but has a complete skeleton of strong linked plates and is densely covered with mobile spines, used in locomotion and defence. It also has small pincer-like organs called "pedicellaria", which are used in defence and, like the spines, may be poisonous.

Above: One of the five arms of a starfish is seen here attacking a sea squirt. Like the sea urchin the starfish is covered with small spines on its upper surface and its arm edges are often set with large defensive plates. The tube feet on the underside are used for grasping prey, this time a member of the tunicate group.

Right: Sea anemones and jellyfish are both members of the cnidaria and therefore both possess stinging cells on their tentacles. Here, a jellyfish is struggling with a sea anemone, but it unclear which is winning.

UNDERWATER FLOWERS

These may look like underwater flowers and the coral reef can be described as a garden but do not be fooled! All these organisms are in fact animals, despite their symmetry and beautiful colours. Most people who enjoy studying rock or tide pools have come across the common red anemone looking rather like a chrysanthemum when it is open, with its tentacles waving in search for food (**Right**). However, it is difficult to imagine that it is the same animal when it is closed and looking like a blob of dark red jelly. Anemones belong to the group cnidaria, which also includes corals and jellyfish. They are the simplest multicellular animals, lacking most of the organ systems that characterise higher animals. Their body is a simple contractible tube and they are, in fact, the equivalent of the coral polyps, being generally sessile and stalked. However, unlike the corals, they are never colonial. Some anemones have longer stalks than others, like the jewel anemone (**Right**), which help to lift the anemone off the sea floor to where it is easier to obtain food. This jewel anemone also shows that the body is like a thin-walled sac with only one opening: the mouth/anus. Plumose anemones also have long stalks and look rather like mushrooms (**Far Right**).

405

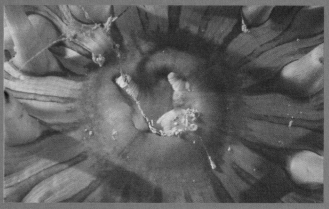

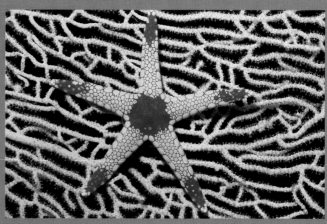

The tentacles of the anemone are covered in sting cells and as they wave in the water, as in the giant dahlia anemone (**Top**), small sea animals touch them. The sting cells paralyse the prey which is then wafted through into the anemone's mouth (**Top Right**).

Everybody loves starfish with their five armed symmetry, virtually unparalleled in the animal kingdom (**Centre**). The familiar starfishes, or sea stars, have five hollow arms linked to an ill-defined central disc (**Centre, Right**). This looks rather like a mouth but in fact the mouth is on the underside, where this starfish is feeding on the surface of the coral. Starfish are mostly scavengers or predators and they grasp their prey by means of tube feet, which are in five rows on their underside. These are small, hollow, walking and feeding organs stretching from the central mouth to the arm tips. Here starfish are seen on sponges. (**Above, Left and Right**). Starfish come in a variety of attractive colours (**Right**) and belong to the group of marine invertebrates called echinoderms, which also include sea lilies, sea urchins, brittlestars and sea cucumbers. As well as five-part symmetry a common feature of the group is an internal skeleton based on calcite, a mineral based on calcium carbonate.

the S EABED

Having visited the fragile world of the inter-tidal region, where terrestrial and marine life seem to go hand in hand, we entered the ocean itself, swimming with an amazing variety of plants and animals, adapted to a truly aquatic environment. Now we will plunge to unknown depths to see what we can discover about the deep, dark world of the seabed; the very floor of the ocean. Here we will find a very strange and wonderful landscape that has more in common with science fiction than with the world that is familiar to us.

Most people are well-acquainted with much of the flora and fauna living in the upper reaches of the open ocean, but the floor of the ocean, particularly in the depths, is often alien ground. The biological communities closely associated with the sea floor, from the inter-tidal zone to the bottom of the deepest ocean trenches, are collectively known as the benthos. A typical benthic community includes scavenging cleaner shrimps, living among the waving tentacles of anemones and feeding on discarded crumbs of food, clams feeding on particles of debris, predatory starfish, crabs, sea urchins, snails and tube worms.

In shallow seas, particularly in regions of high phytoplankton productivity, the benthos is supplied with abundant food and is therefore correspondingly rich in nutrients. The salt concentration can fluctuate at the surface, where rainfall or freshwater from rivers dilutes local areas, but the larger part of the deep ocean has a constant salt content. These are the factors with which bottom dwelling animals must contend and to which they must adapt for survival.

Right: This filter-feeding fluorescent vase sponge lives on the deep reef in the Caribbean.

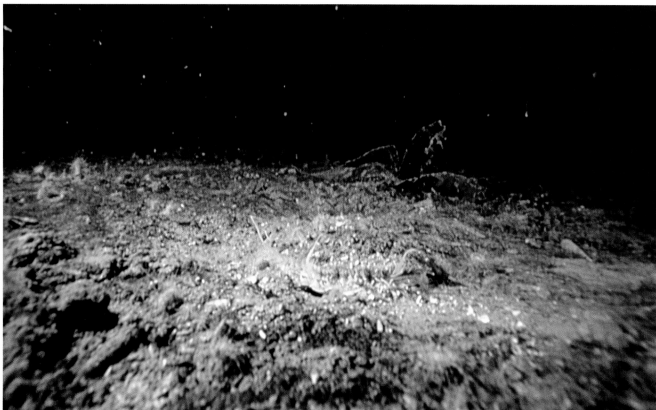

Top: A predatory starfish feeding on mussel beds.

Above: The sea floor is often covered in mud, with few animals in sight.

Above Right: The long legs of the spider crab are ideal for walking across mud.

The composition of the seabed varies from rocks to coarse sediment and thick mud. This determines the type of animals that can be found, hence there are a number of different communities, each adapted to a particular type of substrate. Much of the seabed tends to be covered with silt, soft clays or mud-like oozes made of skeletons of tiny sea animals, so plants and animals must be specially adapted if they are not to sink into the soft seabed. Adaptations include a flattened form, spines, stalks and long legs with hairs. The ooze that covers the vast plains of the seabed can reach several hundred metres thick, and creatures walking on the bottom have to have long legs to avoid stirring it up.

In areas where there are coarse sediments or rocky outcrops, filter feeders such as worms, mussels, sea anemones and sea lilies can survive. Sea lilies use their feathery arms to gather food and several kinds live on the floor of the deep sea in the trenches. Some have roots and stems anchored to the seabed while others with whorls of spikes, called cirri, around their stems, can move by using their arms, dragging their stems behind them.

Some animals associate themselves with another creature to aid their feeding, and thus survival. Brittle stars associate with sea pens or sponges to raise themselves off the sea floor. Climbing off the seabed in this way gives the brittle stars a better chance to collect food. They use their snake-like arms to cling on to their host and then feed on the small creatures and other food particles drifting by. Brittle stars and sea pens are both suspension feeders. They are common bottom dwellers from shallow water to the deep sea, in

oceans all around the world, though they are particularly abundant on the continental shelf and continental slope.

However, the majority of animals that live on the seabed are either scavengers or predators. Scavengers live on any animal or plant matter sinking through the water and are therefore known as deposit feeders. Consequently, the deeper a scavenger lives the less food it is likely to come across because the animals above have eaten more of the available supply. Some scavengers live on the sediment while others burrow into it, feeding on particles and clumps of organic matter that descend from the upper ocean. If large clumps settle on the sea floor, those animals that are mobile, such as shrimps and starfish, gradually move into the area to feed. Fish arrive later, attracted by particles released by the feeding activity of animals that got there first. Because food is so scarce and temperatures so low, most animals living deep in the ocean take a long time to grow and the vast majority are quite small.

Sea cucumbers are slow growing and are found on soft, sandy areas, from shallows to the greatest ocean depths. This mobile echinoderm, from the same group as sea urchins and starfish, leaves tracks from its extra long tube feet as it crawls along the soft seabed.

Living on the rocky seabed in the coastal waters of the Atlantic Ocean and the Mediterranean and Caribbean Seas, is the common octopus. It hides during the day in its rocky lair, coming out at night to look for food, such as crustaceans. It slowly approaches its prey, then pounces, wrapping it between the webbing at the base of its arms. One way for animals to get away quickly in water is by jet propulsion and the octopus uses this method, in common with other molluscs such as clams and squid, which are faster than the octopus because their bodies are permanently streamlined. Squids and octopi when threatened may also eject a cloud of ink to confuse any attackers.

Also associated with the sea floor are, of course, fish. Some are free swimming pelagic fish and have the usual aids to buoyancy associated with fish in the upper layers. This usually involves a swim bladder—a gas-filled sac in the upper part of the body cavity—although truly benthic fish have no aids to buoyancy and spend much of their life in contact with the deep sea floor.

Above: This tiny shrimp is often the first animal to arrive at a new source of food.

Above Left: Sea anemones are common on rocky seabeds.

Left: A brittle star takes advantage of the current of water flowing into this red vase sponge.

Above: Sea cucumbers are sausage-shaped animals with virtually no skeleton. The ring of tentacles around their mouths sweep the surface of the seabed for debris.

Most of the latter are cartilaginous fish, which have softer cartilage skeletons instead of bone, and the characteristic gill slits on either side of the head. Among the cartilaginous fish are the deep sea rays and other flat fish which are typical benthic forms, being dorso-ventrally flattened and very well-camouflaged for life on the seabed. Rays can be more than 200 centimetres (80 inches) in length and cruise the bottom of the seabed hunting for prey such as smaller fish, crustaceans and molluscs. They are much more active than the "sit and wait" predators that comprise the majority of the benthic fauna. A common European flat fish is the flounder—the popular food fish from the seabed.

Among the most common types of predatory fish on the upper slope of the continental shelf are the scorpion fishes; lurking, well-camouflaged creatures that feed largely on small fish and crustaceans, but sometimes feed exclusively on the creeping brittle star. The angler fish is a fearsome batho-pelagic predator with a huge well-armed mouth and long fang-like teeth that are hinged to allow large prey to be engulfed and not escape. Prey are attracted to the waiting jaws by the illuminated stalked lure between the eyes. The wolf fish has large teeth to grip its prey too, and it can bite through crabs, mussels and sea urchins. It is even reputed to be able to bite through a broom stick, though there can't be too many of those on the ocean floor!

Further from land, beyond the shallow continental shelf, the benthic environment can be divided into two zones, the bathyal and abyssal. The bottom is mostly covered by fine sediment and so the animals of the deep ocean floor are generally deposit feeders. The occasional suspension feeder can be found on outcrops of rock but the supply of food is low so these are relatively rare. The creatures of these depths have a very slow growth rate and a long life span.

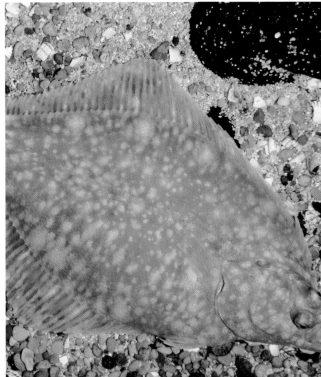

During the last 30 years, experiments with baited traps and cameras have revealed the presence of a previously unsuspected community of highly mobile deep sea scavengers. They range from crustaceans and shrimps, through a variety of fish, to larger sharks many metres in length. All are rapidly attracted to any large food source, apparently over considerable distances. Within a few hours these mobile scavengers arrive at the bait, and reduce it to a well-cleaned skeleton. The small food particles released by this activity in turn attract the much less active scavengers—the echinoderms and molluscs. Active scavengers seem to be highly adapted opportunists that can survive for long periods without food and then rapidly home in on a suitable source when one arrives.

The deep sea still holds many secrets that are gradually being detected. One report from deep water researchers of a huge plume of water led to the discovery of a vast underwater volcanic eruption. They found that more than 300 square kilometres (115 square miles) of the seabed were covered in fresh lava. More fascinating were the huge clouds of white flocculated material, which looked just like snow, being ejected from cracks in the rocks. This snow turned out to be by-products of micro-organisms, and the source was below the rocks. The hot water venting from the cracks contained bacteria, which the researchers were able to culture in

Above Left: The octopus has eight arms covered with suckers to catch its prey.

Top: Many flat fish like this flounder can alter their colouration and markings so as to become almost invisible on the seabed. It is just possible to see its two eyes on the upper surface of its flattened body.

Above: Rat tailed fishes are one of the most successful bentho-pelagic fish, living between the upper waters and the continental shelf. They have a mixed diet derived from food from both the upper and lower regions of the open ocean.

Overleaf: The scorpion fish is found in both tropical and cold waters.

416

Above: The fierce looking wolf fish lurks in crevices and under rocks.

Left: The sampling of the deep sea floor has to be carried out by remote control.

Below Left: A deep sea spider crab moves over a mussel bed in search of food.

Right: Giant tube worms were among the first new animals to be discovered growing near a hydrothermal vent.

the laboratory. Microbiologists were surprised to find that the bacteria were "freezing to death" in the vent water at 55 degrees Celsius (130 degrees Fahrenheit) as they were more used to temperatures in excess of 90 degrees Celsius (195 degrees Fahrenheit). This hotter water must be under the surface itself, and the present theory is that the magma was injected into the rocks from much deeper in the crust, upsetting a sub-surface system and forcing much of the water, along with the bacteria, onto the ocean floor. Now there is the suggestion that there is a new type of ecosystem existing in the deep sub-surface rocks just waiting to be discovered!

One area of the deep sea world that is producing many new discoveries is that of the hydrothermal vents, which were first discovered in 1977, off the Galapagos Islands. Three kilometres (two miles) below the surface a deep sea research vessel came across hydrothermal vents, or hot gushers, where black clouds of super-heated mineral-rich water pour from "chimneys" on the sea floor. Down here on the deepest part of the ocean floor bizarre-looking creatures were being discovered. The temperature of the water was in excess of 400 degrees Celsius (750 degrees Fahrenheit) and within the jet of water were millions of bacteria that were feeding on the chemicals in the water. The bacteria were in turn eaten by giant tube worms. Each was more than three metres (ten feet) in length and these tube worms bore no resemblance to those of the coastal seas. They had no mouth or gut but instead they had red, blood filled tentacles at the tip of their body, which absorbed the bacteria. There were also huge clams, more than three metres (ten feet) long, feeding on the bacteria, while small pink fish and white crabs fed around the worms and clams. Currents in the water carried organic debris from other parts of the sea floor and this was food for many other strange animals, previously unknown to science.

Since this first discovery, many more hydrothermal vents have been explored and every expedition brings back new organisms, not only new species but new genera, new families and even new orders. These gushers are just like geysers on land, where water seeps down through cracks in the crust, getting hotter and hotter. It doesn't boil, because it is under terrific pressure but finally the hot water gushes back up. When it hits the cold water of the ocean it is cooled rapidly and dissolved minerals, including zinc, copper, iron, sulphur compounds and silica, drop onto the ocean floor. The material hardens into chimneys, known as "black smokers" and one of the largest chimneys, extending 50 metres (165 feet) above the sea floor, is nicknamed "Godzilla". There have been many more surprises down at these newly discovered vents. The bacteria around the vents were found to be living inside the clams and worms, breaking down other chemicals into usable food, providing an ecological niche nobody had suspected they could fill. Of all the animals in and around these unique hydrothermal vents, less than ten per cent are found beyond the vents, and, amazingly, more than half the animals were again new to science.

This is a truly endemic community that shows little alliance with the surrounding deep sea and there are many explanations of the great diversity of life that exists around these vents. One theory suggests that it is the unusual conditions down on the deep ocean floor and another suggests that it is the isolation, both in time and space, that has given rise to these strange populations.

Hydrothermal vents are an incredibly difficult place in which to live because of the steep temperature gradient between the hot water venting from the sea floor and the surrounding cold water. The animals have to live within the jet of water and they have enzymes in their cells that can survive the extremely high temperatures and the high concentration of heavy metals. Even a slight change in temperature can kill them instantly. Hydrogen sulphide is highly toxic and must be immobilised quickly, so that it is made safe for the animals to live.

Common marine animals such as the sponges, corals, sea stars and cucumbers are not found in the vents. Many of the animals are so different from any we encounter near the surface that they can be described as living fossils. They have probably lived around the vents, isolated from the rest of the animal world, for millions of years. If the conditions that support life remain the same, there is no need for the animals to change and evolve. Many biologists now believe that the very first organisms on earth were chemosynthetic—that is to say, they were chemical feeders. They also suggest that the vents may well be the best laboratory available for studying how life on the planet actually began.

Above and Above Left: It is risky for filter feeders, such as sea anemones, to live on the seabed—they can become covered with silt, preventing them from feeding.

Sea squirts or tunicates are among the most common animals on the sea floor (**Above Left**). The light bulb tunicate has a transparent coat which gives it a "lit up" appearance (**Above**).

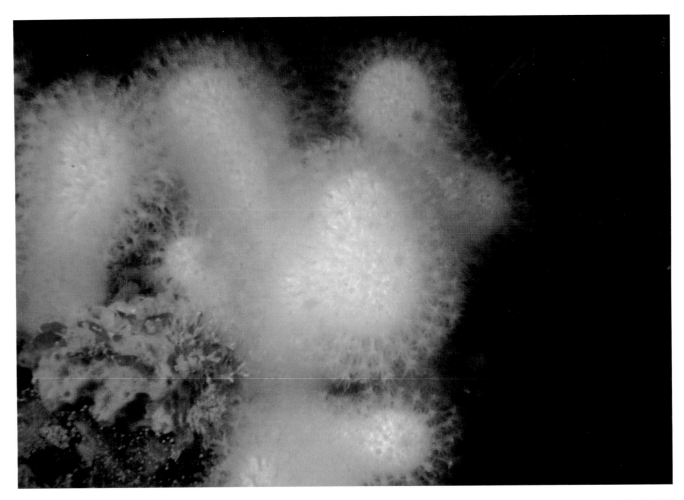

Dead man's fingers is a coral made of rounded, tough masses, which are flesh-coloured when taken out of water. Its has a fuzzy outline underwater caused by numerous colourless polyps each with eight tentacles (**Above Left**). Sea pens are related to the corals. They look a bit like a feather or a quill sticking out of the seabed and they can reach one metre (three feet) in length, with their "feather" sieving the water for food (**Left**).

Sea hares are sea slugs (**Top and Above**) and are related to the tropical nudibranch. They lack the typical shell of a snail, having instead a flat plate and prominent tentacles that look like ears.

Above: Brittle stars are often found congregated together on the sea floor.

Above: Large colourful sunstars are not only found on coral reefs. This sunstar is living in the cold waters around Shetland, Great Britain.

Left: The sea floor is covered with an army of conch shells on the move in search of a new source of food.

Two of the most sought-after seafoods live on the sea floor; the lobster (**Above**) and the edible crab (**Left**).

Above Right: People think of crabs scuttling sideways across the sea floor, but they can also swim.

Right: A scavenging flat worm glides gently over the surface of the mud in search of food.

Above: A fat sea cucumber, with a tough leathery body, lies beside a tiny brown and beige nudibranch.

Sand dollars (**Above Right**) are flat, disc-shaped burrowing sea urchins that live just below the surface of the sand where they eat detritus. Sometimes the tide will lift them from the sand and carry them towards the shore. The five holes in their spiny covering act as spoilers, causing them to tilt and drop to the floor where they promptly bury themselves again. The common sea urchin (**Right**) is much rounder and lives on the sea floor, especially in rocky areas.

Above: Starfish have the ability to re-grow limbs that are broken off. This allows them to repair any damage inflicted by predators who may grab one of their legs.

Above Right: The queen scallop is a common bivalve on the sea floor. Scallops move from place to place by "jumping". They squeeze their shells together, forcing out water and pushing themselves forwards.

Right: Sea spiders look like land spiders, but belong to a group called pycnogonids, and they can have a leg span of 60 centimetres (25 inches). This enables them to stride along without stirring up any clouds of particles. By launching off the seabed, they can swim by bringing their legs up and then sinking down again.

Above: This fan worm has a mucus tube that is completely sunk in the substrate so that the ring of tentacles lie on the seabed.

Left: The top knot, a type of flat fish, lies hidden on the sea floor.

The angler fish waits (**Above Right**), with its lure dangling over its mouth. Its irregular shape can be seen as it swims over the sea floor (**Right**).

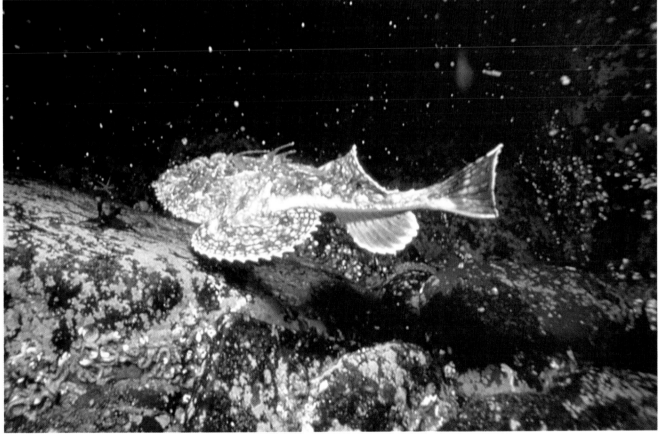

435

Top: The plaice is an economically important fish. Considerable damage is done when trawls are dragged over the seabed to catch plaice and other flat fish.

Above: Gurnards are well adapted to life on the seabed as they have anteriorly placed pelvic fins that have elongated outer rays. Together with prolonged rays in the lower lobe of the tail, these raise them up out of the turbidity on the bottom.

Right: The catfish is a scavenging fish with long whisker-like barbels around its mouth to detect food on the sea floor.

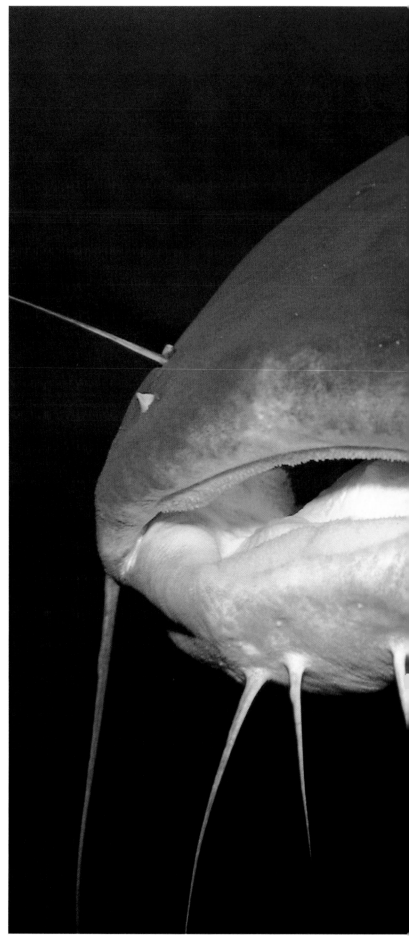

The dogfish is a small shark found in shallow waters (**Top**). It lays its eggs in tough leathery pouches commonly called mermaid's purses (**Above**). Those that are found washed up on the shore attached to seaweeds are usually empty.

Left: The colours of the cuckoo wrasse are variable, but always colourful, ranging from red through to green and blue.

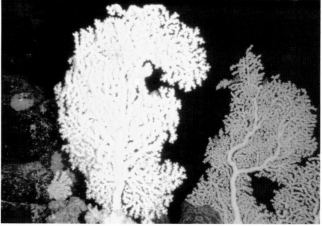

Top: A conger eel investigates a diver.

Above: The golden crab is the largest crustacean on the continental slope of Florida.

Above: A gold-coloured coral grows on pillow lava at a depth of 300 metres (1,000 feet) off the Hawaiian coast. Unlike its relatives growing near the surface it can survive in the dark.

Above: A conger eel investigates a diver.

The community of animals found living around a hydrothermal vent is very different to that found anywhere else on the sea floor. They have adapted to living in the hot mineral waters gushing out of a crack in the sea floor (**Left**). These giant tube worms (**Left, First From Top**) may not look very large but they can grow to lengths of up to three metres (ten feet).

Left, Second From Top: A super-salty brine pool at 800 metres (2,625 feet) is surrounded by a bed of mussels.

Bottom Left: The sea floor is explored using mini rovers with video and laser scanning devices as well as remote-controlled arms to pick up samples.

the **F**UTURE

The earth's continents and islands, atmosphere and oceans are not separate elements, but are interlaced components of a finely-tuned global system. Whatever we do to the land affects the oceans. Changes to the oceans affect the atmosphere and in turn, changes to the atmosphere affect global weather patterns.

More than one-third of the human population lives within 60 kilometres (37 miles) of a coastline. Historically, access to oceans was important for trade, and a safe harbour was a valuable asset. Major cities subsequently grew up around the best harbours and near river estuaries. Cities were built on rivers that provided links to the sea and hence to the major trading routes.

Today, the oceans of the world are under threat from many different sources; not just direct pollution by sewage, rubbish and radioactive waste (though for far too long, the oceans have been used as dumping grounds for all forms of waste) but also from indirect changes. For example, atmospheric pollution is causing global warming, and raises ocean temperatures. On top of this, marine life has to cope with natural changes, such as the El Niño effect that produces an unusually high sea surface temperature over the eastern tropical and equatorial Pacific Ocean, along the coast of North and South America.

Left: Rio de Janeiro is built around one of the world's finest natural harbours.

Right: Rubbish collects on the sea floor where this diver is doing an underwater "litter pick".

Above and Left: Raw sewage can be seen dispersing from an outfall pipe onto a rocky shore (**Above**). In some places, a long outfall pipe (**Left**) takes the sewage out to sea but the sewage tends to be swept back onto the coast by the tides.

Far Left: An unattractive pile of rubbish washed up on a Cornish beach.

445

It is easy to dump something in the water, for the tides and currents quickly wash it out of sight. For centuries, coastal towns and cities have used the oceans as a convenient place to dispose of sewage and general rubbish. As the human population has grown, so has the pollution in the seas.

Most affected by pollution from the land are the shallow coastal waters. Such water is thoroughly mixed by the currents and winds, so the pollution affects all marine life. Polluted water has less effect in the ocean depths. More than half of the world's sewage ends up in the ocean. Some of it is treated, but in the developing parts of the world, and even in some of the better-known Western coastal cities, untreated sewage is dumped. Raw sewage, of course, carries bacteria that contaminate the water, making shellfish unfit to eat and causing illness in swimmers. However, all sewage, whether treated or untreated, is rich in nutrients, especially nitrates and phosphates. Some comes from the organic matter, the rest from soap powders and detergents. When the nutrients enter they water, they cause nutrient enrichment or eutrophication. This, in turn, causes the rapid growth of algae, creating algal blooms over the surface of the water, which affect the marine food web. As the algae die, they are decomposed by millions of bacteria that rob the water of its oxygen. Algal blooms are particularly prevalent in shallow coastal waters or lagoons in which there is little or no current to disperse the pollution. Red tide blooms are caused by an increase in dinoflagellates (marine alga characterised by having two flagella) that can cause shellfish poisoning in humans and mass mortality of clams. Nitrate fertilisers used by farmers also contribute to the problem. These fertilisers are soluble in water, so they drain from the fields and into the rivers, eventually washing into the sea.

Pesticides that are sprayed onto crops and the pest animals that they kill also end up in the sea, becoming incorporated into the food chain. Levels of a pesticide in the water may be very low, but they are taken up into the bodies of plankton which are eaten by fish and other herbivores. Carnivores then eat the herbivores. Each time, the pesticide becomes more concentrated in the animal's tissues. The animals that are worst affected are the top predators—fish eagles, seals and whales—and many thousands of birds and marine mammals have died from pesticide poisoning.

Top: The outfall from a copper mine empties into the sea in Tasmania.

Above: The skeleton of a marine mammal wrapped up in plastic line.

Far Left: An algal bloom in the marina at Newport Beach, California.

Left: During the 1960s, the pesticide DDT was used to kill insect pests. Although it was banned from use a long time ago, it is a particularly persistent chemical that takes a long time to break down, so it is still present in the environment. It has even been found in the bodies of penguins of the Antarctic, carried there by the ocean currents.

THE MEDITERRANEAN

The Mediterranean Sea is virtually enclosed—the only exchange of water takes place through the Straits of Gibraltar. The whole of the Mediterranean is a popular tourist destination and millions of people fly from northern Europe to spend their holidays on its sandy beaches. The coastlines of Spain and Greece, in particular, have seen incredible levels of development. Where there were once picturesque fishing villages, there are now high-rise hotels and holiday apartments. These areas are under considerable environmental pressure. Until recently, most of the sewage was simply pumped straight into the sea where the excess nitrates from sewage and farming caused huge algal blooms in the Adriatic. Meanwhile, fishing increased to supply the tourist industry, putting pressures on the fish stocks, especially tuna. Tourism is not the only problem however. There is a lot of heavy industry along the Mediterranean coastline and this has led to additional pollution from oil, chemical and heavy metal wastes. The poor quality of the water in the late 1980s forced local governments into action. In 1990, countries that bordered the Mediterranean signed the Nicosia Charter. This legislation monitors and regulates discharges into its shallow waters and the quality of the water is now beginning to improve although there is a long way to go, especially with new tourist areas along the North African coast being developed.

Above: High rise hotels line the waterfront in Benidorm, Spain.

Above: Controls on net mesh sizes prevent undersized or immature fish being caught. However, this can cause problems if the breeding females make up most of the larger fish.

Above Left: Over-fishing of cod has prompted Canada and the European Union to suspend cod fishing off the Canadian coast and Iceland has agreed to halve its traditional cod harvest. Hopefully, cod numbers will recover and fishing will be allowed to resume—but this time in a sustainable fashion.

Traditionally, rubbish is either put into landfill sites on land or taken out to sea and dumped. Despite much tighter regulations that now exist in many countries, a vast amount of rubbish still ends up in the oceans. A large proportion of this is plastic. Marine animals often die from either becoming entangled in the plastic or from eating it. The problem with disposing of the waste either on land or in purpose-built incinerators is the cost—it is so much cheaper to sail out to sea and toss the waste over the side. And to make matters worse, it is not just rubbish that is dumped, but also radioactive wastes and toxic mud dredged from rivers and estuaries. Such mud is frequently rich in heavy metals and toxic chemicals.

Industrial processes such as paper making, electricity generation, steel making, mining and oil refining, all produce a large volume of waste water that is emptied either into local rivers or the sea. Pollution from these sources causes a lot of damage in shallow coastal zones because the polluted water tends to get carried back towards the coast by the tides. Unfortunately, these coastal zones tend to be where the largest concentrations of fish can be found. Enclosed bodies of water such as Chesapeake Bay, the Baltic, and the Caspian and Mediterranean seas are particularly vulnerable. The high density of people living around these shores simply magnifies the problem.

In 1989, the world fish catch reached an all time high of 100 million tonnes. Since that time, the overall harvest has fallen back slightly to 97 million, with decreases in some oceans and increases in others. In 1950, the catch was less than 30 million tonnes. We have now got to a point where most seas and oceans have reached or even passed the maximum harvest that they can offer without diminishing future harvests. It seems that over-fishing, almost to the point of extinction in some cases, has become an accepted fisheries management philosophy.

Fisheries are a natural resource. They must be managed on a sustainable level as they can easily be reduced, but not so easily increased. Fisheries can only be protected if harvests are kept below the level at which fish are taken faster than they can be replenished. Many countries have made attempts to regulate their fishing industry; the European Union, for example, has substantially reduced its fishing fleet over the years, but still the fishing stocks are under threat. Hard decisions have to be made when there is a stark choice between jobs and conservation. Sometimes, however, there is no choice and fishing simply has to be banned. In the 1960s, North Sea herring stocks collapsed completely, and fishing for herring had to be banned for ten years to allow the population to recover. It has now done so and the harvest has remained constant at a level just below that last seen during the 1950s.

Over the years, enormous government subsidies have created fishing fleets composed of modern, highly-efficient trawlers. Massive factory boats use high-tech electronics to track the shoals, so more fish are caught per run than before. As more trawlers compete for fewer fish, intense competition is generated between fleets. In an attempt to control fishing in European waters, the European Union operates a quota system, though with limited success, for as catches decline, fishermen are tempted to break the rules. The last resort is to prohibit fishing in spawning grounds and nursery areas or to have closed seasons in which no fishing may take place.

Subsidies cause their own special problems. The larger industrialised nations, such as the USA, Japan and the European Union, give other nations huge subsidies to buy new fishing vessels. The aim of

the subsidy is to protect the donor nation's fish supply. The European Union gives approximately US$1.4 billion each year in fish subsidies. Fish subsidies of US$162 million allowed Argentina to completely modernise its fishing fleet. The stocks of hake off the Argentinian coast were once the most productive in the world, but, thanks to the over-fishing made possible by the new equipment, the fisheries are now subject to a complete fishing ban. Up to 70 per cent of the world's fish stocks need urgent intervention to prevent a further decline. Among the most at threat are the fish stocks of Southeast Asia, the China Sea, Argentina, the Southern Ocean, Eastern Canada, Iceland and Great Britain.

Over-fishing also affects other marine life. Fish are a critical part of the food chain and over-fishing means other animals go hungry. The puffin, for example, relies solely on sand eels. If sand eels decrease, so do the puffins. The problem is compounded by the damage being done to coastal habitats such as salt marshes, estuaries and mangrove swamps. These are the spawning and nursery areas for many types of fish. In Italy, for example, 95 per cent of the coastal wetlands have now been drained with the obvious consequence that the fish now have fewer places to spawn and so stocks have become more limited.

Drift nets are many kilometres in length, and hang in the water to catch pelagic fish. Unfortunately, these nets are not selective and many animals become trapped within them, dying for no purpose whatsoever. Quite often the nets are abandoned and left to float free in the oceans, becoming a death trap for even more animals. It

is impossible to estimate just how many dolphins, sharks, turtles and other animals have died in these nets.

Our lack of concern for the value of species is clearly demonstrated by the global slaughter of sharks. Shark dorsal fins are used to make a simple bowl of soup, but no other part of the shark is used. Many sharks are highly vulnerable to overfishing because they grow slowly, mature late and produce a small number of young. They can become over-fished rapidly and, once depleted, can require decades to recover. Some sharks congregate in special nursery grounds in shallow waters to give birth or "pup". At these times, the large numbers of pregnant females and juveniles are particularly vulnerable to fishing. As a consequence, many species of shark are classed as endangered. In contrast, whaling has raised a great deal of public interest. For centuries, whales have been hunted for their meat, blubber and oil. The blubber was burned to provide light, the meat was eaten and the oil used in lamps. In the early days, whales were hunted from small boats with hand thrown harpoons but, in

Above: A turtle has become entangled in a net. Some of the latest nets are turtle friendly so they can escape.

Above Left: More than half of the mangrove swamps in tropical areas have been lost.

Top: The catch of shark from a single fishing trip in the Persian Gulf.

Above: The *Braer* ran aground off the coast of Shetland in January 1993.

Above Right: A sperm whale is towed up the slipway of a whaling station in Iceland in 1977.

Right: Whale watching is very popular, demonstrating that whales have a considerable worth if allowed to live. By protecting whales, countries can have a guaranteed revenue stream for the foreseeable future.

1864, the Norwegians developed an explosive harpoon that enabled many more whales to be caught. After the First World War, factory ships were used to support the whaling fleet, which allowed the whalers to stay at sea for months on end and catch hundreds of whales. In 1900, there were 2.2 million whales but fewer than 600,000 remain today. The great whales, including the blue, humpback, right, sperm and grey, were the largest and most profitable species so were hunted the hardest. As the great whales became rarer, the whalers started to hunt the smaller species, so eventually all species were severely reduced in numbers. As the whales of the northern oceans all but disappeared, the whalers turned their attention to the rich pickings of the Southern Ocean.

In 1946, the International Whaling Commission was established to meet each year and agree on quotas. Since most of the members were whaling nations, little was done to enforce the quotas and scientific advice was frequently ignored. Whaling reached its peak in 1961, when a total of 60,000 whales were slaughtered. Pressure to ban whaling increased, led by groups such as Greenpeace. At last, in 1972, the UN voted for a ten-year moratorium on whaling, but it was not until 1985 that the IWC agreed to halt the commercial hunting of whales. There were exemptions made for scientific research and for the Inuits, for whom whale meat is a traditional food. Stocks have now begun to improve to such an extent that the traditional whaling nations are lobbying for whaling to begin again. Each year, the IWC has to defend it decision to continue the ban. These nations may get their own way, but public opinion seems to be firmly on the side of the whale.

Every day, thousands of tankers sail along the world's shipping lanes, carrying oil from the oil terminals to refineries in the industrial countries. There have been some major oil spills in recent years, including the *Exxon Valdez* in Alaska, the *Braer* in Shetland, Great Britain, and, most recently, the *Erica* off the coast of Brittany, France. The oil forms a slick over the surface of the ocean and is washed up onto shorelines, coating the sand and rocks with a layer of oil. This kills the invertebrate animals straight away as they cannot breathe. Sea birds, seals and otters also become coated in oil. The oil clogs the feathers of the birds and they are unable to fly. As they try to clean their feathers they swallow the oil and die. Sea otters lose their protective insulation and freeze to death.

CORAL HEALTH

Coral reefs are under threat from a number of sources, including global warming, El Niño effects, pollution and over-exploitation by humans. A small rise in sea temperature causes coral polyps to expel the zooxanthellae algae, to the point at which a coral with no algae at all loses its colour. It turns white and is said to be "bleached". Coral polyps can survive short periods without their algae, while the zooxanthellae become re-established as the water temperature drops. However, if higher temperatures persist for a longer period, the coral dies. In recent years, large-scale bleaching has been reported on the reefs around the Maldives and on a smaller scale in the Caribbean and on the Great Barrier Reef. Fortunately, records indicate that bleaching events have occurred in the past and that, in the long-term, reefs do recover.

It is not just global warming that is a threat to coral reefs. Slash and burn agriculture in undeveloped tropical regions denudes the

Above: A reef-walk on Heron Island, Great Barrier Reef, is guided by a marine biologist. There is a fine balance between the benefits of educating visitors about the reef and the damage the tourists may do while they are exploring.

Above Right: Some of the more popular reefs are visited by too many divers.

landscape of trees. Without tree roots to absorb water and hold the soil together, heavy rainfall washes excessive amounts of soil into rivers and out to sea. The silt, and a consequent reduction in sunlight, chokes coral polyps. Algae feed on the silt and reproduction of algae gets out of balance with the natural checks placed on it by the reef. Sewage, chemicals and agricultural fertilizers washed into the sea all have a similar effect on coral reefs.

Island nations often have few natural resources, and a coral reef may be an important tourist attraction. Unfortunately, the rise in the popularity of diving and the larger number of tourists now undertaking long-haul holidays means that some of the best known coral reefs are being damaged by hoards of tourists. Too many divers, the anchors of dive and snorkel boats and the litter and sewage that inevitably accompany them, all contribute to a loss of diversity on the reefs. It is important that tourism is controlled and managed in a sustainable way. If the coral reefs lose their interest, the visitors will go elsewhere.

Attractive fish are also collected by locals to be sold to the aquarium trade. Sometimes, the locals use cyanide to paralyse the fish so that they can be caught easily, but this can kill invertebrates in the area. Marine aquarists can help to protect reefs by insisting that the fish they buy have been bred in captivity.

When visiting a coral reef, it is important to remember that the coral is a living entity. Just stepping on a coral can destroy years of growth. Tourists want souvenirs of their holiday, but they have to remember where the attractive conch shell or coral fragment came from.

Marine conservation organisations have set up reef monitoring schemes called "Reef Watch". People are asked to complete questionnaires so that the organisation can monitor events such as coral bleaching.

Traditionally, spilled oil is treated with detergent to cause it to break up. But detergents can often do more harm than good. The best approach is now thought to be to use skimmers to lift the oil from the surface of the water and to use hot water sprays on oil that has reached the beach. However, it may well be best to simply leave the clean-up to nature. Oil is biodegradable and is broken down by bacteria so, in time, it will disperse naturally. However, it is certainly an eyesore and people who use the beaches need them to be free of oil.

Many of the world's tanker spills could have been prevented by making oil tankers with a double hull so that if the outer hull was breached the inner one would have remained intact. Of course, this adds quite a lot to the cost of a new tanker and single-hulled tankers will still be sailing for many years to come.

While oil spills continue to hit the headlines, other leaks into the oceans are often ignored, but each year huge amounts of oil enter

Top: Fossil fuels are being burnt in ever increasing quantities, releasing carbon dioxide into the atmosphere.

Above: As the earth warms up, the polar glaciers are retreating.

Right: This guillemot was one of the thousands of sea birds smothered in oil when the *Sea Empress* was wrecked off the Welsh coast in 1996.

the oceans in the forms of small discharges and leaks from tankers; for example there is little thought for the environment when ships wash out their tanks.

Marine life is acclimatised to living in water of a certain temperature. Because of the large volume of water, temperatures below the surface remain remarkably constant. However, a sudden change in the temperature of the water can affect whole communities. Many tropical and arctic marine species live at temperatures close to their upper lethal limits. Slow growing tropical coral reef systems are highly dependent on narrow temperature ranges. A rise in water temperatures of just two to four degrees Celsius (four to eight degrees Fahrenheit) above normal can cause the death of vast marine ecosystems. Sea turtles and salt water crocodile populations are also affected, for the temperature at which their eggs are incubated determines their sex.

Above: During the construction of a new port at Moreton Bay, Australia, a buffer zone was built between the port facility and the nearby mangrove swamp.

Above Left: Seal Rock in California is part of the Monterey marine sanctuary. The shore is frequented by sea lions, harbour seals and cormorants.

Visitors are allowed to watch some of the loggerhead turtles laying eggs on Mon Repos beach, Australia (**Far Left**). Researchers have to relocate some of the eggs after the female has finished, since beach erosion following a cyclone now prevents the turtles climbing above the high tide mark (**Left**).

In the oceans, warming by both the El Niño phenomena and by greenhouse gases affects global ecosystems. The earth absorbs heat energy from the sun and much of it is radiated back to space, but some is trapped in the atmosphere by gases such as carbon dioxide—the so-called greenhouse gases. Without these gases, the temperature on earth would be minus 18 degrees Celsius (zero degrees Fahrenheit), so we do need them. However, since the Industrial Revolution more than 200 years ago, the levels of carbon dioxide and methane have increased as a result of the burning of fossil fuels and the clearance of forests. As a result, global temperatures have risen by 0.5 degrees Celsius (about one degree Fahrenheit). This may not seem much, but it is already having a significant effect on global climate patterns. If the present trends continue, it is expected that average temperatures could increase by as much as three degrees Celsius (about six degrees Fahrenheit) within the next 50 or so years. This would result in glaciers and the polar ice caps melting, causing a rise in mean sea level. Warmer water occupies a larger volume than cool water so, as the temperature of the oceans increase, the volume would increase too. This, combined with ice melt, would create devastating floods in low-lying areas. It would affect places such as Bangladesh and the Netherlands, while islands like the Maldives would be totally submerged. The average height of waves would increase, and storms would become more frequent and severe. Rainfall patterns over continental land areas would change, and the major ocean currents, such as the Gulf Stream, might well be affected. Global warming is already with us, as major breaks in the ice sheets of Antarctica have recently demonstrated.

In the stratosphere, chemicals called CFCs (chlorofluorocarbons) are damaging the protective ozone layer. These are the chemicals once used in aerosol sprays, expanded polystyrene and fire extinguishers. Although they are now banned, they have a very long lifetime and their damaging effects will be felt for several more decades. As the ozone thins, increased levels of biologically-damaging UV-B radiation reaches the surface of the earth. This adversely affects many marine organisms and the entire marine food web. Phytoplankton populations are reduced, the DNA of marine species can be damaged and fish populations are again diminished.

So what can we do to protect our oceans? The very best marine habitats can be given special protection and many countries have well-established marine national parks in which both the coast and the inshore ocean are given total protection. In the USA, there are 12 such parks, with more being planned. In Australia, the beaches where turtles lay their eggs are given particularly strong protection. During the breeding season the beaches are off-limits to the public, who are only allowed to watch the turtles under strictly controlled viewing conditions. In Greece, there has been some conflict between tourism and turtles. The lights of hotels and discos distracted the turtles as they emerged from the sea on their traditional beaches and holidaymakers disturbed their nests. Now, as a result of education and publicity, there is much better control and it is quite common for flights to and from the islands to be delayed in order not to disturb a turtle. In fact, the turtles are proving to be one of the best local tourist attractions.

Sand dunes are easily damaged. The eggs of birds are so well camouflaged that they can be easily broken by people as they walk the dunes. The best way to protect sand dunes is to build a boardwalk, which allows people to cross the area more easily and get a good view of the habitat.

Estuarine environments require more careful control. These areas are often the favoured development sites for the petrochemical industry, as they have good access to both the sea and river. Unfortunately, such wetlands are often important feeding grounds for wading birds and act as fish nursery areas. Careful planning of development can help to minimise damage to the environment. For example, the building of a buffer zone between the industrial area and the estuary will isolate any pollution and help to manage storm waters. In some places, it has proved possible to re-establish the original habitat. Many nations have realised that clearing their mangrove swamps was a mistake. Not only were the swamps important fish nurseries, but they also protected the coast from storm surges. Some mangroves were cleared for new port facilities but, more recently, the main reason for their clearance has been to build

Above Left: Mangrove restoration in Indonesia.

Left: Two manatees swim along a tidal river in Florida.

prawn lagoons. These are ponds in which large numbers of shrimps and prawns are raised for sale to the catering industry. However, the lagoons provided no protection for the coast. Nowadays, the remaining mangroves are being given protection while replanting is being carried out along the coastlines most at risk from typhoons and cyclones.

As well as protecting habitats, it is often necessary to protect individual species.

One species receiving special attention is the manatee. Manatees are slow moving, herbivorous sea mammals that are found in the swamps, estuaries and rivers of Florida, the Carribean and South America. They eat aquatic plants and can consume 10-15 per cent of their body weight daily in vegetation. They have no natural enemies, and it is believed they can live for 60 years or more. However, as a result of habitat loss, there are less than 2,600 manatees remaining in the USA. In addition, manatees are often killed when they collide with watercraft, are crushed in canal locks and flood control machinery or ingest fishing lines and litter. A management plan is now in place, which involves studying the manatees to learn about their behaviour and distribution. It will require the implementation of speed limits for boats, educating the public about the manatee and buying up habitat for the creation of sanctuaries.

The great frozen continent of Antarctica is one of the last true wildernesses of the world. It still relatively unspoilt, despite the presence of research stations and increasing numbers of visitors. However, keeping this continent pristine is very difficult. The extremely cold conditions mean that any pollution or environmental damage persists for a long time. Just a footprint can remain in the snow for years and an oil or chemical spill could devastate huge areas. One of the greatest threats to the continent comes from the huge mineral reserves that lie beneath the ice. In order to prevent mining companies moving into the region, nations with a base on the continent have signed the Antarctic Treaty. This has given both the continent and the Southern Ocean protection from mining and other industrial activity for 50 years and has prohibited activities such as whaling.

What is the future for Antarctica? Life on earth as we know it today is only the blink of an eye in geological terms. To us Antarctica is a frozen waste, but it is known that it was once a green and fertile continent, unlike the Arctic. Some even speculate that it may have been the site of the lost continent of Atlantis. Certainly there are treasures and mysteries locked up in the ice and it would be wonderful if we could dare to hope that the nations of the world, for once, could truly pull together to preserve and maintain this beautiful and extraordinary continent, which can teach us so much. It is the last unspoilt continent on Earth and its future hangs in the balance. Many people believe that the best way to protect Antarctica would be to make it the first World Park where there was no development and the unique plant and animal life found there could continue to live safely. Antarctica is too precious to spoil!

You cannot have read this chapter without initially feeling a profound sense of sadness at the devastation caused by humans to the oceans of the world. Hopefully too, you will have felt that we all have a deep responsibility to do something to reverse the pollution, the over-fishing, the extinction of species, the disruption of delicate food chains and human greed. As you read of conservation measures you would have begun to see some hope, that if we all pull together we can save our seas. The United Nations Conference on Environment and Development, also known as the "Earth Summit", which was held in Rio de Janeiro in 1992, saw politicians, environmental experts and many others gathered together to discuss the challenge that humanity faces in the 21st century. Namely that we must find a way to live without using up all the resources that future generations will need to live their lives. We can no longer live on our little islands and forget that the rest of the world exists: now we must put in global effort to save our blue planet.

The ability of our planet and its inhabitants to survive into the next millennium will depend on how we care for the air, land and sea, for the Earth as a whole is truly more valuable than the sum of its parts. The future of the oceans rests in our own hands.

World Park status would give protection to all of
Antarctica's wildlife.

Above: The sediment dredged from shipping channels is taken out to sea and dumped. This sediment often contains high concentrations of heavy metals.

Above Left: Seas and oceans can be polluted in any number of ways. There are many wrecked ships lying at the bottom of the sea and, in most cases, the cargo sank with the ship. Here, the cargo is one of bottles containing acids that could contaminate the surrounding water.

Left: During storms, cargo can be lost from ships. The coastguard has to deal with containers that get washed up on the shore.

Right: The leachates from landfill sites can enter watercourses. This landfill site in Scotland lies right beside an estuary and the leachate is draining straight into the water.

Above: The incoming tide is bringing in a dense foam, probably from a spillage of detergent further along the coast.

Surfers have long demonstrated against the discharge of sewage into the sea (**Left**). "Surfers Against Sewage" is a campaign group that raises awareness of this issue (**Right**). Surfers spend a lot of time in the water, and are prone to contract infections caused by water-borne bacteria.

Above Left: A filthy black effluent from a Spanish building site pours into the Mediterranean Sea.

Above: Pulp mills use a lot of water in the treatment processes. This mill on Vancouver Island, Canada, treats the waste water before it is emptied back into sea, but not all mills have such stringent controls.

Water around bathing beaches has to be regularly tested for the presence of bacteria (**Far Left**). The European Union has strict limits on the number of bacteria that may be present. Beaches that have water with low levels of bacteria may display a blue flag, showing that it is suitable for public use (**Right**). In contrast, this estuary has been contaminated with raw sewage and people are not allowed to swim in the water (**Left**).

An oil spill close to a coastline, especially one of ecological importance, can have a devastating effect on the wildlife. Sea birds, such as the shag (**Right**) are particularly vulnerable. The oil slick may be blown onto the shore where it contaminates holiday beaches. The popular beach at Tenby, Wales was completey covered in oil following the *Sea Empress* oil spill (**Above**).

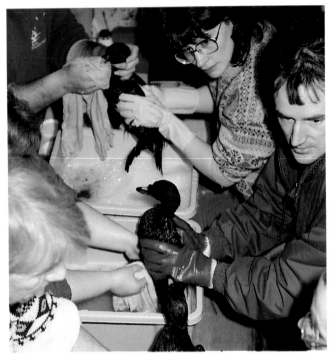

There is little that workers can do with the oil, other than push it into
one area where it can be sucked up (**Top**). Oil birds are taken to rescue
centres where teams of volunteers clean the oil from the feathers.
Unfortunately the survival rate is not very high (**Above**). It may take ten
or more years for the coast to undergo a full recovery. The *Exxon Valdez*
spill took place near the oil terminal at Valdez, Alaska (**Right**). The tanker
at the terminal is surrounded by a barrier to prevent accidental spills of oil
contaminating the water. Sometimes ships are surrounded by a barrier of
bubbles which prevents pollution of the water (**Below**).

Left: A factory fishing ship and its support fleet of trawlers can stay at sea for many months. The support ships unload their catch onto the factory ship which carries out the processing and the freezing of the fish.

Far Left: Shark fins hang on display in a shop in Singapore. They are the vital ingredient of shark fin soup. However, the rest of the shark is thrown away.

Above: The actions of Greenpeace volunteers have raised the public's awareness of issues such as whale hunting.

Above Left: Whale meat is still eaten in the Faroes. Each year there is an organised slaughter of pilot whales, which are driven into the harbour and killed on the shore.

Above: The clearance of tropical rainforest leads to soil erosion. The soil is carried away in streams and rivers and is eventually dumped in the sea.

Above Right: Local fishermen use dynamite to kill fish, rather than nets. This diver is holding a victim of dynamite fishing.

Right: Tourists are often tempted to buy shell souvenirs and it is difficult to know which shells come from licensed farms where the molluscs are raised for their shells and which come straight from the reef.

Far Right: Dynamite devastates the structure of the reef.

Silt covers the coral animals, clogs their polyps and kills them. Coral reefs are thus threatened by development. New hotels are being built on many of the coral islands of the Great Barrier reef (**Left**) and off the United Arab Emirates coast (**Above**). Great care has to be taken that the hotel does not pollute the water with its sewage and other wastes. However, there have been some positive actions taken to protect the reefs. Many reefs have installed permanent moorings for the dive boats so that anchors do not drag on the reef (**Right**). In some areas artificial reefs are being created by sinking surplus army vehicles and naval boats. These wrecks are quickly colonised by corals and sponges, as seen in Marshall Islands (**Top**).

Top: There is competition for beach space between the beach goers and bird life.

Above: The coastal habitats of estuaries and mangrove are under great pressure as these are areas that are favoured for port and tourist developments. A large expanse of mangrove was ripped up for a new oil terminal and refinery in Queensland.

Right: Some coasts have undergone massive development, as seen on the Gold Coast of Australia where the Surfers Paradise resort is very popular.

Estuaries are often dredged and used as marinas. The mudflats are
replaced by deep permanent water and are subject to pollution from
spillage and urban run-off.

Above: Some shorelines are "tidied up" and the natural beach replaced by rock-lined public areas which are far less wildlife-friendly.

Above Right: Even digging for fishing bait on tidal sea grass flats can damage the sea grass and bring anaerobic mud from deep under the surface to the top.

Right: Damage to fragile habitats such as sand dunes can be prevented by providing a board walk. This board walk in Florida gives people easy access to the national seashore near Pensacola.

The effects of global warming would be widespread. The rise in carbon dioxide, a greenhouse gas, levels is due in part to massive deforestation throughout the tropics especially Southeast Asia (**Above Left**). The increased global temperatures will cause of the ice at the ice caps to melt (**Left**). Already there are signs of some of the ice sheets in Antarctica breaking up (**Above**). This will lead in an increase in wave height. The Southern Ocean has the greatest average wave height of seven metres (21 feet) and this will get higher. There will also be more stormy weather (**Right**). In recent years, the number of level three and four hurricanes (**Overleaf**) have increased, leading to more extensive damage.

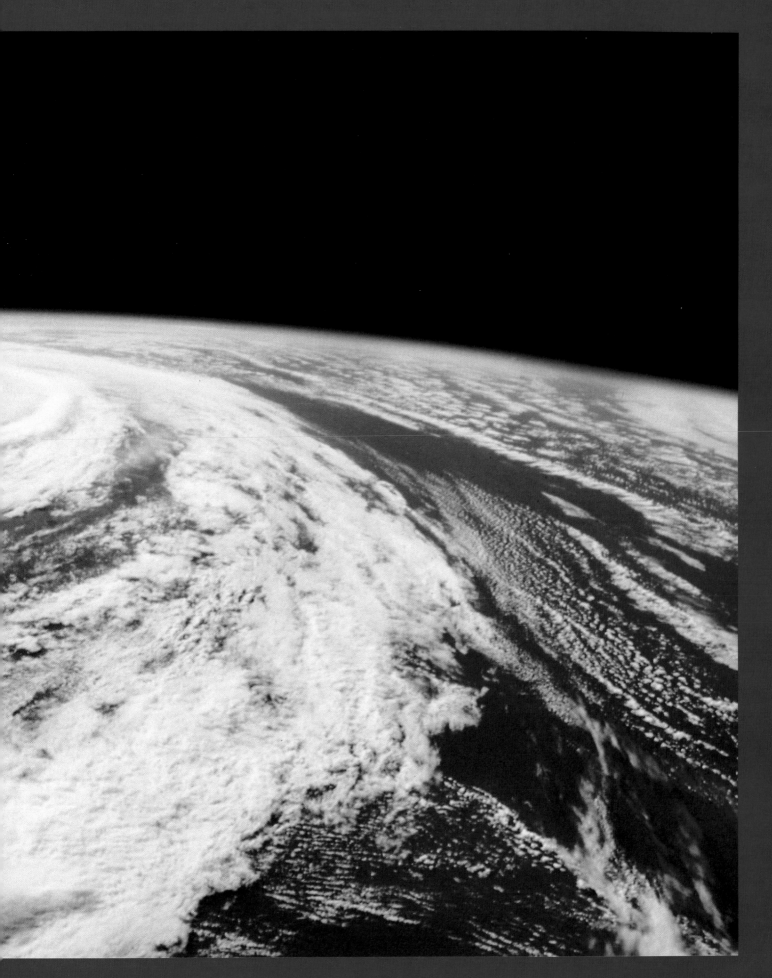

Island nations such as the Maldives lie just a metre (three feet) or so above sea level and will disappear if the sea level increases as predicted. Changes are already taking place. Five years ago, these palm trees on a beach in the Maldives (**Above**) were several metres from the edge of the sea. Increased wave height (**Left**) has eroded the sandy beach, so that it has almost disappeared and rocks have been imported to protect what remains of the beach (**Right**).

Watching sea life is a popular pastime. Visitors are regularly taken by boat to Bass Island, Scotland, where there is a large gannet colony (**Above**). People can also watch whales (**Left**) and seals from the coast (**Below**).

Above: Port Campbell National Park in Victoria, Australia, protects the stunning coastline where erosion has produced features such as rock arches, blow holes and stacks.

These mangrove swamps (**Above**) and sandy beaches (**Left**) lie along the protected coastline of the Cobourg Peninsula in Gurig National Park, Australia.

Top: The Galapagos Islands with their unique community of plants and animals have been classified as a world heritage site.

The National Trust in Britain has purchased thousands of kilometres of coastline and now protects some of the best-known heritage coasts such as Flamborough in Yorkshire (**Left**) and the Seven Sisters in Sussex (**Above**).

Left: The Monterey Bay National Marine Sanctuary stretches from just north of San Francisco, south along the entire Big Sur coast. The sanctuary extends more than 70 kilometres (45 miles) out to sea. Ocean currents and undersea terrain make the sanctuary waters a rich environment that supports a diversity of marine habitats and species. A deep water canyon (one of the largest and deepest of its kind) cuts through Monterey Bay. The canyon is home to an array of deep water animals, living in the waters of the canyon and on its walls and the sea floor.

Left: The stunning Rialto Beach in Washington State with its rock pinnacles.

Whales (**Above**) and dolphins (**Right**) are probably the two best-known symbols of ocean life that need to be conserved for future generations. The well-being of the oceans is essential for all life on earth and the oceans' great natural resources as well as their beauty (**Left**) must be preserved for future generations.

BIKINI ATOLL—
BACK FROM THE DEAD

In 1946, Bikini Atoll was selected as the test site for atomic tests. These tests, code-named "Operation Crossroads" were designed to investigate the effect of atomic bombs on warships. A fleet of surplus American warships, together with captured Japanese and German ships were fuelled and armed as if they were in a state of war-readiness. Then they were moored in circles at ground zero—the point of detonation. The first bomb donated 150 metres (500 feet) above the water, sinking a number of lighter ships including the Japanese cruiser *Sakawa* and the destroyer *Lampson*. The next test involved the first detonation of an atomic bomb underwater. This bomb sent a cloud of radioactive spray into the air and created huge waves which crashed into the target ships. Most were sunk including the aircraft carrier *Saratoga* and the Japanese battleship *Negato*.

In 1954, Bikini was the site of the world's first detonation of a hydrogen bomb. This bomb proved to be far more powerful than expected and it vapourised one of the islands, throwing millions of tonnes of radioactive coral dust into the atmosphere. The hydrogen bomb made headlines around the world and one of the spin-offs was a design of swim suit named the "bikini".

The wrecks of the destroyers *Lampson* (**Top and Above Right**) and *Anderson* (**Above**) are covered with both soft and hard corals as well as sponges. A myriad of small animals live amongst the wreckage.

The *Saratoga* was probably the most famous US aircraft carrier of the Second World War (**Top**). Now the carrier lies upright in 55 metres (180 feet) of water, her anti-aircraft guns intact on either side (**Above**).

The Japanese battleship *Nagato* lies upturned on the sea bed (**Below**). The superstructure has been colonised by sponges and whip corals. Fish such as large sting rays peer from hiding places, while grey reef and white tip sharks watch from in the shadows.

Overleaf: There is a healthy population of grey reef sharks in these waters, seen here in a feeding frenzy.

The radioactive contamination made Bikini Atoll uninhabitable for 40 years. But now the level of radiation has declined to a level that it is considered safe for people to visit the atoll again, so long as they do not eat anything grown locally or swallow a mouthful of mud.

This coral reef has remained untouched during this time, giving biologists the opportunity to see how animals have colonised the wrecked ships. Although a major environmental disaster at the time, it did not take long for the wrecks to become colonized by fish, the structure of the ships providing an imitation of the holes that fish look for in a real reef. Sponges and other encrusting forms came next, using the wrecks to grow in an advantageous position in the current. Hard corals took longer to become established, growing on the shallower parts of a wreck first, where there was more sunlight.

FISHING

The greatest catches of fish come from the temperate and sub-polar continental shelves of the northern hemisphere and from the upwelling areas of the great, open, oceanic fisheries. Although there are some 20,000 species of fish in the oceans of the world, only a small proportion are of commercial value. In temperate waters large numbers are caught but relatively few species, while in the tropics the reverse is the case. Commercial fishing, predominately by trawling, is dominated by the developed nations, while the developing countries rely on fishing as their main source of animal protein in a diet otherwise deficient in this respect, primarily using traditional methods to catch their fish.

The world market has some striking regional variations, and in the developed world there has been a marked shift over the last few decades from fresh to frozen products, and also to high value canned products such as salmon and tuna. However, in many tropical markets, with their less well-developed handling, marketing and storage facilities, fresh fish and traditional dried and salted products are mostly found.

This harbour front in the Canaries (**Above**) has fishermen selling their fresh fish, just caught from their small fishing boats, in contrast to the modern fish market in Holland with its plastic boxes for packaging (**Below**).

Above: Trap fishing is a common method of catching fish and is shown here on the Kenyan coast, where the incoming tide brings the fish into the trap. The fish are prevented from escaping when the door of the trap closes as the water recedes.

Left: This small boat in the Gulf of Oman may employ traditional fishing methods, such as hand lining, but it plays a part in the commercial tuna fishing industry. Commercial fish can be roughly divided into two classes; those which swim near the surface like herrings, mackerel, sprats and tunny, and the deep sea fish like cod, haddock and the flat fish, which live on or near the sea bottom. Whereas the former are usually caught with drift or set nets, the latter are fished with trawls, lines and seines, which are large vertical fishing nets whose ends are brought together and hauled.

Below Left: Spear fishing is widely practised by Arctic races and here we see an Inuit cutting up salmon which has been caught by traditional methods and with respect for the environment. This is an example of sustainable fishing where only enough is caught for the needs of the people, not for commercial gains or for recreational reasons. Bows and arrows are used by some tribes in the Amazon to catch fish and in Siberia a hook and line is used through a hole in the ice.

Above: In parts of Africa and Asia a tapering net is cast, with a large number of weights round the edge and the fish are netted when the tide comes in, as seen in this photograph of two boys in the Philippines throwing a net.

Right: Here we see herring being pumped from a purse-seiner. This factory trawler is a far cry from the sustainable fishing of primitive people who fish to survive.

Below: Modern trawlers are highly sophisticated, with ultrasonic equipment to locate shoals of fish and technology to process the catch as well.

Bottom: Fish are not the only catch netted in parts of the world. Here we find spider crabs caught in a gill net, usually used for catching fish by their gills.

INDEX

INDEX continued

ABOUT THE PHOTOGRAPHERS

Jeff Collett has been a scuba diver for 16 years and interested in underwater photography for the last eight years. He has recently returned from the Gulf area where his long-term residency in the Middle East provided him with both the opportunity for excellent local diving and a central base from which to explore other top diving destinations. Apart from diving extensively in the UAE and Oman his diving interests have taken him to Sipadan, the Philippines, Maldives, Indonesia, Cyprus, Malta and the Caribbean. He is never happier than on a long shallow dive on a Maldivian Island house-reef, accompanied by his regular buddy, shooting away at the bizarre creatures and extravagant sights of the underwater world.

Dr Ben Hextall (top centre) graduated with a PhD in Marine Biology from the University of Liverpool, specialising in community ecology. He is an experienced diver who, for over 18 years has dived in many locations worldwide. He

is an accomplished underwater photographer with an extensive photo library. In 1993 he co-authored a dive guide book, *Dive Isle of Man*, and is also co-author of a comprehensive identification guide to the marine life of the British Isles and north west Europe. Dr Hextal is also director of OceanWeb a successful website design management company specialising in yachting and marine projects.

Wayne Lawler is a freelance photographer based in Australia. He is particularly interested in bird photography, especially water birds and their habitats. He takes a keen interest in environmental issues and tries to show what people are doing to their environment through his photography.

John Liddiard (top right) is a freelance photographer and writer specialising in underwater photography. He likes to spend as much time as possible underwater, both in tropical seas and at home in UK waters. John describes himself as a fanatical diver with an impossible ambition to dive everywhere. Top of his list of unfulfilled ambitions is to dive under Antarctic ice, but the closest he has come to this is diving with the penguins at Bristol Zoo.

Robert Pickett (top left) Works as a professional wildlife photographer, which presents him with may obstacles. Taking pictures that vary from the smallest microscopic subjects to the largest animals and plants of land, sea and air, he is passionate about his work. After taking many land based images, an opportunity presented itself to take up scuba diving, and another extension for photography. The animals and plants that inhabit the world never cease to amaze him and he hopes that through his images he can capture some of the beauty that he sees. Robert now runs, with his partner Justine, Papilio Natural History & Travel Library, where they also have the work of other outstanding photographers, who have also contributed to this project.

ACKNOWLEDGEMENTS

The publisher wishes to thank the following for supplying the photography for this book: Jones/Shimlock/© Wild Images Ltd/RSPCA Photolibrary for front and back cover and pages 6 (cut out), 66 (cut out), 293, 294-295, 296-297, 298-299, 300-301, 302-303 (all), 304-305 (both), 316, 330 (top), 331, 332-333, 335 (bottom), 336-337, 338-339, 342, 343, 352-353, 354-355, 356-357 (all), 358-359 (all), 366 (bottom), 366-367 (main), 368, 369, 370-371 and 442 (cut out); Papilio/Arun Madisetti for pages 2, 7, 24 (top), 25, 29 (middle), 40 (top), 50-51, 162 (cut out), 173 (top), 238 (top), 275, 276-277, 289 (top), 322 (top), 348 (bottom) and 381 (bottom); © Ecoscene/Mark Caney for pages 4, 40 (bottom left), 53, 106 (top), 186, 188, 202 (top), 224-225, 236, 243, 248, 272-273, 284, 340 (top), 443, 464 (top) and 501; © Ecoscene/John Liddiard for pages 6 (top bar), 22 (top), 30 (bottom left), 42 (top), 45, 52 (top), 66 (top bar), 156 (bottom), 162 (top bar), 174 (top and middle), 175, 180, 184 (top and bottom), 189 (top and bottom right), 190 (top left and right and middle), 196, 197 (top and bottom), 202 (bottom), 203, 204-205, 206-207, 208 (top), 227 (bottom), 228 (top and bottom), 229 (top and bottom), 230 (top bar), 231, 232-233, 234, 238 (bottom), 239, 247 (top), 256 (left and bottom right), 265 (top and bottom), 266, 267 (bottom), 268-269, 270 (bottom), 271, 274 (bottom), 281 (top and bottom), 288 (top), 289 (bottom), 292 (top), 306, 310, 314 (bottom), 317 (top), 318 (top), 319, 327 (top), 341 (top and bottom), 347 (left), 350-351, 361, 364, 376 (bottom), 380 (bottom right), 403, 406 (top left and right), 410 (top and bottom), 412 (top), 413, 415 (top left), 418 (top), 423, 424 (top and bottom), 425 (top and bottom), 426, 427 (top), 428 (bottom), 431 (top), 433 (top and bottom), 434 (bottom), 435 (bottom), 439 (top), 440 (top), 442 (top bar), 455, 502 (top and bottom), 503 (top left and right, middle) and 504-505; Genesis Space Photo Library for pages 6 (bottom), 12-13, 55 (bottom), p68 (bottom) and 488-489; Papilio/Claudio Corradini for pages 8, 246 (top), 247 (bottom), 393 (bottom), 395 and 493; © Sylvia Corday Photo Library Ltd/Matthew D Harris for pages 9, 177, 226 (bottom), 235 and 479 (top); © Sylvia Corday Photo Library Ltd/Dr Ben Hextall for pages 10, 20 (top), 21, 23, 27 (bottom right), 31 (top right and bottom right), 36-37, 41 (bottom), 62, 70-71, 101, 145 (top), 146 (bottom), 147 (middle), 148-149 (main), 150 (top and bottom left and right), 151 (top right), 152-153, 157, 170-171, 185 (bottom left), 187 (inset), 192-193 (main), 242, 244-245, 262-263, 270 (top), 278-279, 280 (top), 401, 404 (bottom), 405, 414, 416-417, 420, 422, 428 (top), 429 (top), 431 (bottom), 432, 434 (top), 435 (top), 436 (top), 438-439 (main) and 439 (bottom); © Sylvia Corday Photo Library Ltd/Bunny Warren for pages 11, 107 (top), 179 (bottom), 215 (top), 463 and 487 (top); © Sylvia Corday Photo Library Ltd for page 14 (top); Papilio/Marcus Walden for page 14 (bottom); Papilio/Melvyn Lawes for page 15 (top); Papilio/Stephen Coyne for pages 15 (bottom) and 29 (bottom); © Ecoscene/Wayne Lawler for pages 16 (top), 18 (bottom), 43 (top), 78-79, 80, 84-85, 91, 92-93, 94-95, 100 (bottom left), 112-113 (main), 116-117, 118, 120 (top), 121, 124, 125 (top and bottom right), 127 (top), 216 (top bar), 408 (top bar), 459, 460 (top), 476 (left), 480 (top), 482-483, 484, 485 (top), 486 (top) and 494 (top and bottom); © Ecoscene/Andrew Brown for pages 16 (bottom), 67, 76, 86 (right), 444, 445 (top) and 495 (middle); © Sylvia Corday Photo Library Ltd/Jonathan Smith for pages 17, 54, 87, 89, 138, 181 and 223 (bottom right); © Ecoscene/Robert Nichol for pages 18 (bottom), 19 (top), 122-123, 220 (top), 453 (bottom), 457, 470 (left) and 492 (top); Papilio/Robert Pickett for pages 19 (bottom), 24 (bottom), 26, 27 (bottom left), 29 (top right), 30 (top left and top right), 38 (top), 39 (top and bottom), 40 (bottom right), 42 (bottom), 104, 168 (top), 173 (middle), 174 (bottom), 187 (main), 193 (right), 195, 208 (bottom), 226 (bottom), 292 (bottom), 307, 324, 326 (bottom), 334-335 (main), 335 (top), 360 (top), 362 (inset), 374 (bottom), 375 (top), 376 (bottom), 377 (bottom), 404 (top and middle), 411, 415 (top right) and 436 (bottom); Papilio/Robert and Justine Pickett for pages 20 (bottom), 30 (bottom right), 31 (top left), 250-251, 253 (bottom), 257, 261, 274 (top), 286-287, 291, 325, 330 (middle), 344-345, 348 (top), 380 (bottom centre) and 394 (middle); Papilio/Dinah Aldam for pages 22 (bottom), 33 (top), 172, 198-199, 230 (cut out),

255 (bottom), 264, 280 (bottom), 323, 327 (bottom), 349, 362 (main) and 429 (bottom); © Dr Rod Preston-Mafham/Premaphotos Wildlife for pages 27 (top left), 90, 100 (top left), 144 (bottom left and right), 146 (top), 147 (top), 149 (top), 151 (bottom), 154 (top), 165 (bottom), 166-167 and 213 (bottom); © Ecoscene/Elgar Hay for pages 27 (top right), 34, 60 (top), 322 (bottom), 412 (bottom) and 430; © Ecoscene/Jeff Collett for pages 28, 31 (bottom left), 32 (top and bottom), 33 (top), 35, 41 (top), 46 (top and bottom), 47, 48-49, 222 (top), 223 (top), 230 (bottom), 237 (top and bottom), 240, 241, 255 (top), 256 (top right), 258, 260, 267 (top), 285, 288 (bottom), 290 (top and bottom), 308-309, 311 (top and bottom), 312-313, 314 (top), 315 (top and bottom), 317 (bottom), 318 (bottom), 320-321, 326 (top), 328-329, 330 (bottom), 346, 347 (right), 365, 366 (top), 372 (top), 373, 374 (top), 375 (bottom), 377 (top), 378, 379 (top and bottom left and right), 380 (top and bottom left), 381 (top), 382, 383 (top and bottom), 384 (top and bottom right and left), 385 (top and bottom right and left), 386-387 (main and inset), 388-389, 390-391, 398 (top and bottom), 399, 400 (top and bottom), 402, 406 (middle left and right, bottom left and right) and 407; © Ecoscene/James Marchington for page 29 (top left); © Ecoscene/Sally Morgan for pages 38 (bottom), 44 (bottom), 81 (bottom), 97, 103 (top), 129 (bottom right), 144 (top and bottom centre), 154 (bottom), 182, 183, 246 (bottom), 392, 393 (top), 442 (bottom), 446, 450, 454, 458 (top), 468 (top right) and 496-497; Martin Edge/© Wild Images Ltd/RSPCA Photolibrary for pages 43 (bottom), 252, 340 (bottom) and 363; Papilio/Bryan Knox for pages 44 (top), 96, 100 (top right), 158 (bottom), 159 (top), 218 (top left), 447 (bottom left) and 495 (top); © Sylvia Corday Photo Library Ltd/Chris North for pages 52 (bottom), 163, 223 (bottom left) and 475; © Sylvia Corday Photo Library Ltd/John Parker for pages 55 (top), 178 and 259 (bottom); © Sylvia Corday Photo Library Ltd/Dizzy de Silva for pages 56 (top) and 218 (bottom); OAR/National Undersea Research Program (NURP) for pages 56 (bottom), 60 (bottom), 168 (middle), 168 (bottom) photographer M Youngbluth, 173 (bottom), 190 (bottom), 189 (bottom right) photographer R Griswold, 221 (bottom) photographer R Wickland, 415 (bottom) photographer S Ross, 416 (middle left), 416 (bottom left) photographer I MacDonald, 416 (bottom right) photographer C Van Dover, 427 (bottom) photographer G Wenz, 440 (bottom left) photographer R Cooper, 440 (bottom right), 441 (top) photographer P Rona, 441 (upper middle), 441 (lower middle) photographer J Brooks, and 441 (bottom); © Sylvia Corday Photo Library Ltd/John Farmar for pages 57, 98 (top), 99 (top right), 160 (bottom) and 169; © Sylvia Corday Photo Library Ltd/Jill Swainson for pages 58-59; Papilio/Michael Maconachie for pages 61 and 155 (top); © Sylvia Corday Photo Library Ltd/Robert Whistler for page 63; Papilio/Dennis Johnson for pages 64 (top), 98 (bottom) and 160 (top); © Sylvia Corday Photo Library Ltd/Edwin Baker for page 64 (bottom); © Ecoscene/Simon Grove for pages 65, 102 (bottom) and 155 (bottom); © Ecoscene/Chinch Gryniewicz for pages 68 (top left and top right), 72 (bottom), 73, 77, 81 (top), 100 (bottom right), 103 (bottom left), 113 (bottom), 114-115 (main), 115 (top), 129 (bottom left), 456 (top), 464 (bottom), 467 (top and bottom), 472 (top and middle), 485 (bottom), 500 (bottom), 506 (top) and 507 (top); © Ecoscene for pages 69, 164, 194, 209, 397, 474 (bottom left), 477, 490 (top), 491 and 503 (bottom); © Ecoscene/Anthony Cooper for pages 72 (top), 75 (left), 130-131 (main) and 492 (bottom right); © Ecoscene/Genevieve Leaper for pages 74, 134-135, 213 (top) and 465 (top); © Sylvia Corday Photo Library Ltd/Humphrey Evans for page 75 (right); © Sylvia Corday Photo Library Ltd/Chris Taylor for pages 82-83; © Ken Preston-Mafham/Premaphotos Wildlife for pages 86 (left), 103 (bottom right), 120 (bottom right), 141 (main), 145 (bottom), 147 (bottom), 149 (bottom) and 156 (top); © Ecoscene/Ken Ayres for pages 88 and 165 (top); © Ecoscene/Barry Hughes for pages 98 (middle), 115 (bottom), 131 (right), 158 (middle) and

159 (middle left and right, and bottom); © Ecoscene/Robin Williams for page 99 (top left); © Sylvia Corday Photo Library Ltd/Chris Parker for pages 99 (bottom) and 179 (top); © Ecoscene/Richard Glover for page 102 (top); © Ecoscene/Peter Hulme for pages 106 (bottom), 129 (top) and 474 (top); © Ecoscene/Luc Hosten for pages 107 (bottom), 216 (middle right) and 227 (bottom); © Ecoscene/Graham Neden for pages 108-109; Papilio/Dorothy Burrows for pages 110 and 158 (top); © Ecoscene/Ian Beames for pages 111 (top), 113 (top) and 133 (bottom); © Ecoscene/Alexandra Jones for pages 111 (middle), 119, 120 (bottom left), 222 (bottom right), 447 (top), 458 (bottom left and right), 480 (bottom) and 492 (bottom left); © Ecoscene/Andy Hibbert for page 111 (bottom); © Ecoscene/Alan Towse for pages 125 (bottom left), 161, 191 (bottom), 220 (bottom), 221 (top), 472 (bottom), 476 (bottom right) and 478; © Sylvia Corday Photo Library Ltd/James de Bounevalle for pages 126 and 460 (bottom); © Sylvia Corday Photo Library Ltd/S K Tiwan for page 127 (bottom); © Ecoscene/Nick Hawkes for pages 128-129 (main); © Ecoscene/John Farmar for pages 129 (upper middle), 132, 133 (top), 214-215 (main), 219 (bottom), 222 (bottom left), 394 (top) and 474 (bottom right); © Ecoscene/Kay Hart for page 129 (lower middle); © Ecoscene/E J Bent for pages 136-137 and 396; © Sylvia Corday Photo Library Ltd/Nick Smyth for pages 139 and 490 (bottom); © Sylvia Corday Photo Library Ltd/Les Gibbon for pages 140-141 (main), 219 (top), 495 (bottom) and 500 (top); © Ecoscene/Stephen Coyne for pages 142-143 and 215 (bottom); © Mark Preston-Mafham/Premaphotos Wildlife for page 151 (top left); © Sylvia Corday Photo Library Ltd/Nick Rains for pages 176, 282-283 and 360 (bottom); Papilio/Björn Backe for page 185 (top); © Ecoscene/Robin Redfern for page 185 (bottom right); © Ecoscene/Robert Baldwin for pages 191 (top), 227 (middle), 451, 452 (top), 479 (bottom left) and 507 (middle); Rick Rosenthal/© Wild Images Ltd/RSPCA Photolibrary for pages 200-201 and 407 (cut out); © Ecoscene/Mike Whittle for pages 210-211 and 253 (top); © Ecoscene/Peter Dannett for pages 212-213 (main); Papilio/David Smith for pages 213 (middle) and 218 (top right); © Sylvia Corday Photo Library Ltd/John Penam for page 215 (middle); © Ecoscene/Quentin Bates for pages 216 (top and middle left), 217, 449 (left and right), 506 (bottom), 508 (middle and bottom) and 508-509 (top); © Ecoscene/Baldwin, Wilson, West for page 254; © Sylvia Corday Photo Library Ltd/Graeme Goldin for page 259 (top); © Sylvia Corday Photo Library Ltd/John Jones for page 372 (bottom); Papilio/Sylvain Olivera for page 394 (bottom); © Cliff Nelson ARPS/Premaphotos Wildlife for page 409; © Ecoscene/Joel Creed for page 421; Papilio/Clive Druett for pages 436-437 (main); © Ecoscene/Paul Ferraby for page 445 (bottom); © Ecoscene/Erik Schaffer for pages 447 (bottom right), 468 (top left) and 472-473 (main); © Ecoscene/Jim Winkley for pages 448, 480-481 (main) and 498-499; © Ecoscene/Kieran Murray for page 452 (bottom); © Ecoscene/Tom Ennis for pages 453 (top) and 507 (bottom); © Sylvia Corday Photo Library Ltd/Carola Holmberg for pages 456 (bottom) and 486 (bottom); © Ecoscene/Rod Gill for page 465 (bottom); © Sylvia Corday Photo Library Ltd/Nigel Rolstone for pages 466 and 468 (bottom left); © Ecoscene/Ian Harwood for pages 468 (bottom right) and 469; © Ecoscene/Michael Cuthbert for pages 470-471 (main); © Ecoscene/Martha Collard for pages 476 (top right) and 479 (bottom right); © Sylvia Corday Photo Library Ltd/Geoffrey Taunton for page 487 (bottom); © Ecoscene/Michael Gore for page 508 (top).